AF725077

SEX WORK IN SOUTHEAST ASIA

Southeast Asian sex workers are stereotypically understood as passive victims of the political economy, and submissive to Western men. The advent of HIV/AIDS only compounds this image, as sex workers come to represent the victims of, and vectors for, a deadly virus. From sensationalist stories of Bangkok bar girls to United Nations conventions on global sex trafficking, the media, activists and academics alike condemn uneven processes of economic development and uneven relations of power in the sale of sex. While the goal of such commentary may be to transform the economic and social relationships that make sex tourism a reality, the stereotypical images it produces often bear little resemblance to the everyday experience of sex work.

Sex Work in Southeast Asia is a cultural critique of HIV/AIDS prevention programmes targetting sex tourism industries in Southeast Asia. Drawing on experiences of community-based organizations, national governments and emerging opinions from the international prostitutes rights movement, it highlights how feminist and postcolonial politics shape practices of global AIDS prevention. Women in the industry until now have rarely been included in the dialogue, so in this book they are given voice to reveal their own conceptions of working in the sex industry. By juxtaposing practical, contemporary issues of AIDS prevention with current theories of subjectivity and identity, *Sex Work in Southeast Asia* posits a new place for a speaking sex worker subject. This book will be vital up-to-date research for scholars in cultural, political, social and urban geography, as well as in development and gender studies.

Lisa Law is Australian Research Council Postdoctoral Fellow at the Gender Relations Project, Research School of Pacific and Asian Studies, Australian National University, Canberra.

ROUTLEDGE PACIFIC RIM GEOGRAPHIES
Series Editors: John Connell, Lily Kong, and John Lea

1 LANDSCAPES OF GLOBALIZATION
Human geographies of economic change in the Philippines
Philip Kelly

2 SEX WORK IN SOUTHEAST ASIA
The place of desire in a time of AIDS
Lisa Law

SEX WORK IN SOUTHEAST ASIA

The place of desire in a time of AIDS

Lisa Law

London and New York

First published 2000
by Routledge
11 New Fetter Lane, London EC4P 4EE

Simultaneously published in the USA and Canada
by Routledge
29 West 35th Street, New York, NY 10001

Routledge is an imprint of the Taylor & Francis Group

© 2000 Lisa Law

Typeset in Times by Florence Production Ltd, Stoodleigh, Devon.

Printed and bound in Great Britain by
TJ International Ltd, Padstow, Cornwall

All rights reserved. No part of this book may be reprinted or reproduced or utilized in any form or by any electronic, mechanical, or other means, now known or hereafter invented, including photocopying and recording, or in any information storage or retrieval system, without permission in writing from the publishers.

British Library Cataloguing in Publication Data
A catalogue record for this book is available
from the British Library

Library of Congress Cataloguing in Publication Data
Law, Lisa, 1967–
Sex work in Southeast Asia: the place of desire
in a time of AIDS/Lisa Law.
p. cm. (Routledge Pacific rim geographies; 2)
Revision of thesis (doctoral—Dept. of Human Geography,
Australian National University, 1996).
Includes bibliographical references and index.
1. Prostitution—Asia, Southeastern. 2. Sex tourism—Asia,
Southeastern. 3. Sex-oriented businesses—Asia, Southeastern.
4. AIDS (Disease)—Social aspects—Asia, Southeastern,
5. AIDS (Disease)—Asia, Southeastern—Prevention.
I. Title. II. Series.
HQ241.A5.L38 2000
306.74′2′0959—dc21 99–087581

ISBN 0–415–21805–5 hbk
ISBN 0–415–24669–5 pbk

CONTENTS

ILLUSTRATIONS

PLATES

FIGURES

TABLES

ACKNOWLEDGEMENTS

Many individuals and institutions have made writing this book possible. First of all, John Connell and Simon Whitmore have provided the encouragement and logistical support that brings *Sex Work in Southeast Asia* into print. The book draws on my doctoral dissertation, which I completed in the Department of Human Geography, Australian National University (ANU) in 1996. The Gender Relations Project (GRP), ANU supported additional research in Indonesia and Thailand in early 1999. For providing stimulating conversation and careful guidance during my PhD, I thank Doug Porter, Margaret Jolly, Alison Murray and Peter Rimmer. For friendship and critical conversations, I am indebted to Glenn Banks, Sally Oliphant, Karl Benediktsson, Juan Pulhin, Lisa Drummond, Chienwei Yeh and Gilly Robson. I am particularly indebted to Rosemary Wiss for critical insights and a delightful sense of humour. My work has also benefited from the stimulating intellectual environment provided by Ien Ang, Mandy Thomas and Elaine Lally at the University of Western Sydney, and by Margaret Jolly and Shelley Mallett of the GRP. Conversations with Steve Pile have proved an endless source of inspiration, and encouragement from Kathie Gibson has been truly appreciated.

In the Philippines, many people and organizations helped me think through the very complex issues of sex work and HIV/AIDS, and I have attempted to represent their ideas and insights here with care and respect. These people include: Resil Mojares and Linda Alburo at the University of San Carlos; Ilya Abellanosa of the City Health Department in Cebu; Geoff Manthey of the World Health Organisation; and members of non-government organizations such as Kabalikat, Talikala, Visayas Primary Health Care and Services, WEDPRO and the Women's Resource Center of Cebu. In Bali, Indonesia Made Suprapta and Christopher Dureau of the Indonesia HIV/AIDS and STD Prevention and Care Project gave me the opportunity to think through the issue of culturally sensitive AIDS prevention programmes. They also introduced me to the Citera Usada Foundation, who kindly explained their AIDS education project for Balinese beach boys. In Bangkok, Thailand I received the generous assistance of Sharalyn Jordan and Surang Chanyam of Empower.

My greatest debt of gratitude is to the women employed in Cebu City's sex industry. These women were both my friends and informants, and allowed me the space to interrogate my own conceptions of prostitution, HIV/AIDS and sex tourism. They also helped me think through issues of women's agency, sexuality and cultural difference. Most importantly, they helped me realize that the increasingly global discourses that represented them as an intelligible group were strategic, partial and loaded with ethical considerations, and I have attempted to make this an explicit part of my writing. I wish them all the best in their pursuit of a 'better life'.

Finally, I thank Michael Bird, who has spent more time in go-go bars than he would ever have imagined, and continuously brings happiness and a sense of balance to my life.

The following articles have been revised for inclusion here and appear with permission. Chapter 3 is a revised version of Cartographies of desire: mapping Cebu's sex industry in communal/plural, *Journal of Transnational and Crosscultural Studies* 6(1): 87–105. Chapter 4 originally appeared as Dancing on the bar: sex, money and the uneasy politics of third space, in Steve Pile and Michael Keith (eds), *Geographies of Resistance* (London and New York: Routledge, 1997), pp. 107–23. Chapter 6 is a revised version of Local autonomy, national policy and global imperatives: sex work and HIV/AIDS in Cebu City, *Philippines in Asia Pacific Viewpoint* 39(1): 53–71. Permission has also been granted for my article, A matter of choice: discourses on prostitution in the Philippines, in Lenore Manderson and Margaret Jolly (eds), *Sites of Desire/Economies of Pleasure: Sexualities in Asia and the Pacific* (Chicago: University of Chicago Press, 1997), pp. 233–61, which was my first attempt at thinking through issues of agency and choice in the sex industry.

SERIES PREFACE

The Pacific Rim is the world's most dynamic region, confronting diverse economic, political and sociocultural changes of different orders of magnitude. The region has been open to, and the impetus behind, forces of globalization. The transcending of borders has taken place in political, economic and sociocultural spheres. Economically, transnational corporations have emerged as strong contenders in the world of capital, establishing and maintaining business networks in varied parts of the world, not least Asian business networks. Culturally, globalizing forces have given rise to transculturation, the interchange of cultural elements and the breaking down of cultural borders. Politically, states have had to be concerned about the maintenance of the 'nation' in the midst of larger cross-national forces. Yet, at the same time, indigenous practices have been asserted through patterns of localization, with local (often ethnic) identities becoming increasingly important as geographical phenomena.

The Pacific Rim is actor and driver in all these processes, as globalization and localization are constantly being reconstructed in new and imaginative ways, and as modernity is transformed and differently imagined. This series seeks to provide a platform for the analysis of such developments in Asia (or parts of Asia), as well as island states within the Pacific Ocean region. It will include scrutiny of issues such as:

- environmental change (e.g. resource management, sustainable urban development, deforestation and global warming);
- urban transformations (e.g. city cultures, urban management and planning, migration, housing, global cities, the informal city, tourism and urban change;
- culture and identity (e.g. religion, music, food, fashion, gender);
- local and regional change (e.g. globalization and local autonomy, entrepreneurs, transnational corporations, the informal sector, rural development).

All these, and more, form the empirical situations within which the global and local intersect, and relate to the core themes of culture, environment, urbanization and geographical change that this series seeks to address.

The books in this series will resonate with diverse themes in as much as the authors reflect a variety of backgrounds and experiences, offering rich insights both from within and outside the region. We hope the diversity will be rewarding and appealing to a wide interdisciplinary audience.

John Connell
Department of Geography
University of Sydney
email connelljohn@hotmail.com

Lily Kong
Department of Geography
National University of Singapore
email lilykong@nus.edu.sg

John Lea
Department of Architecture
Planning and Allied Arts
email lea@arch.usyd.edu.au

LIST OF ABBREVIATIONS

AFAO	Australian Federation of AIDS Organizations
AIDS	acquired immune deficiency syndrome
ANU	Australian National University
AusAID	Australian Agency for International Development
CATW	Coalition Against the Traffic in Women
CSW	commercial sex worker
DOH	Department of Health
ECPAT	End Child Prostitution in Asian Tourism
GABRIELA	General Assembly Binding Women for Reform, Integrity, Equality, Leadership and Action
GRP	Gender Relations Project
HIV	Human Immunodeficiency Virus
IV	intravenous
KAPB	knowledge, attitude, practice and belief
MCDP	Metro Cebu Development Plan
MEPZ	Mactan Exporting Processing Zone
NGO	non-government organization
NIC	Newly Industrializing Country
NSWP	Network of Sex Worker Projects
PO	people's organization
STD	sexually transmitted disease
USAID	United States Agency for International Development
VPHCS	Visayas Primary Health Care and Services
WEDPRO	Women's Education, Development, Productivity and Research Organisation
WHISPER	Women Hurt in Systems of Prostitution Engaged in Revolt
WHO	World Health Organisation
WRCC	Women's Resource Center of Cebu

INTRODUCTION

In the late 1980s Asia was designated the new epicentre of HIV/AIDS, and a diverse assortment of commercial sex activities were grouped together under the rubric of 'sex work'. Sexual encounters negotiating different sexualities and genders, different classes, races and ethnicities, went under the epidemiological microscope, and became the subject matter of health professionals, activists, academics and government agencies. 'Sex workers', as they are increasingly called, thus emerged in the global discourse of HIV/AIDS – and in a manner conditioned by ideological debate. Representations of Southeast Asian sex workers as victims of the global political economy, as casualties of tourism development as well as prey to (and vectors for) a deadly virus became particularly prevalent in the 1990s, gaining a new and transnational coherence.

While AIDS research in Southeast Asia has revealed multiple forms of paid sex, as well as different levels of foreign involvement, AIDS prevention activities have continued to place a strategic emphasis on paid sexual relations between Western men and Asian women. The major preoccupation of these projects and programmes is the extent to which contact with foreign men implicates women in the local spread of HIV, and how to prevent the contaminating influence of foreign sexual relations. Prevention activities are influenced by popular and academic texts, where portrayals of HIV infection highlight the crucial role of sex tourism in bringing HIV to countries in the region and how this predicament reproduces long traditions of foreign exploitation. The sex worker's body becomes conflated with nation, invaded by foreign powers and infected with a fatal virus. Such fetishized representations permeate AIDS education activities, and are interrogated in this book.

It is important to acknowledge how foreign sex industries are implicated in the spread of HIV, but the tropes of male power and Western dominance in the region have marginalized the voices of sex industry workers themselves. Dichotomous models of power permeate AIDS prevention discourse – that is, the colonized/colonizer binary of nationalist discourse as well as feminist discourses of masculine oppression and female subordination – reinscribing stereotypical images of Asian women as passive and exploited victims. Indeed, the sex worker community debated among AIDS experts, bureaucrats and activists has more to do with anti-colonial and feminist politics than with the everyday experience of sex work. While a few studies have been devoted to the experiences of sex workers in Asia, and some sex workers have moved up through the ranks of community organizations, it is more often the case that they are spoken for. Although I myself cannot claim spokesperson status, it is possible to interrogate why this voice is not heard.

Sex Work in Southeast Asia examines the emergence of a sex worker subject through AIDS education activities in this dynamic Asian region. By contextualizing the complexity of anti-colonial and feminist politics, it investigates the assumptions that motivate intervention into the lives of sex workers and implications for understanding the agency of individuals who engage in this (usually temporary) profession. It begins with three stories from the Philippines, Thailand and Indonesia that bring to light different issues raised in the book. These stories contrast the diverse forms of commercial sex with the simplicity of representations used to describe them, and ground an understanding of the sex industry in the place I contend it matters most: the lives of its participants.

Lives and community: Cebu City, Philippines

> The first time I visited Alumnos it appeared a destitute squatter community. The tiny houses were cobbled together bits of wood and metal, had dirt floors, and very few had running water. Despite their size, about 5 × 5 metres on average, these structures housed large extended families. Most had a kitchen/eating area, and one or two bedrooms to accommodate several sleepers on straw mats. While a few residents were optimistic enough to upgrade their homes by putting down cement floors or building extensions, the uncertainty of land tenure meant that capital investments would be lost if the landowner reclaimed the property. Many could simply not afford to. There was also the problem that Alumnos, a seaside community, was prone to flooding during the rainy season. This meant rock-hopping for several months a year, as well as fighting off fierce swarms of mosquitoes.
>
> The surfaces and edges of Alumnos seemed to brighten with each visit I made. People in the community smiled and called out as they recognized me, and I had a favourite place to buy soft drinks and sweets. I knew enough families to see lives and relationships constrained by finances but which were about more than just poverty. People in Alumnos had problems but they also had hopes, dreams and interesting stories to tell. I had come to

know these families through their daughters, wives and mothers – women who worked in Cebu's tourist bars geared to white, Western men.

One afternoon I visited Alumnos with my friend and landlord Virgie.[1] We met when she was a go-go dancer at one of the local bars, although she had recently resigned after starting a relationship with an Australian tourist. While Virgie was delighted to be off the evening schedule, it was clear she missed the sociality of the job. The women of Alumnos were regular employees of Cebu's bars, Virgie knew many of them and she was pleased about the opportunity to catch up on everyone's news. We'd arrived with her 3-year old son Joey, who had developed a nasty cough over the past week. Virgie was also eager to take him to Margaret, a practising hilot skilled in traditional massage/healing.

As Margaret massaged Joey's chest, the conversation quickly shifted to life in Negros – an island directly west of Cebu. Both she and Virgie grew up in small Negros villages, and their nostalgic conversation drifted from fields of sugarcane to missed loved ones and food. Margaret migrated to Cebu City at the close of World War II. During the occupation (1942 to 1944), the Japanese burned many of the villages and crops in Negros, destroying the most basic recourse to livelihood. Although the Americans liberated Negros by 1945, Margaret and her husband moved to Cebu in search of greener pastures. While Cebu had not turned out they way they had expected, Margaret and Virgie shared the dream of a better life for their children.

After Margaret had finished Joey's massage, I asked if she had photos of her children and (now deceased) husband. She entered the house while we sat on the verandah, and returned with two cardboard boxes of miscellaneous bits of paper and snapshots. At the bottom of one of the boxes were two framed 8 × 10 photos of her eldest daughters who were now living in the United States. Before displaying them, she explained that they had run off to Angeles in 1975 without her permission, and found jobs in the bars for American servicemen. She pulled out the photo of the oldest daughter, who married an American and now lives in California with three children. Margaret declared she had made a success of her life, unlike the younger daughter who married an American but divorced shortly after and now lives in Chicago alone. The only family photo of the younger daughter was an enlarged snapshot of her in a bikini, and someone had scratched a hole in the vicinity of her crotch. A telling reminder that although Alumnos bar women are respected for the money they bring in for their families, it is always remembered how they earned it.

> For Margaret, bar work was a good source of income and an investment in the future if her daughters were lucky enough to marry an American. Memories of the Americans liberating the Philippines from the Japanese remain vivid for Margaret and many of her neighbours. If women in Alumnos worked in Angeles and found an American husband, they were on their way to a better life.

Images of sex tourism in the Philippines have abounded for the past three decades. Whether in the red light districts of Manila or Cebu, or at the American military bases in Angeles and Olongapo, images of middle-aged Western men violating adolescent Filipinas have inspired intensive critiques of American colonialism, cultural imperialism and the commodification of Filipino sexuality. While the important interpretive frames provided by anti-colonial accounts of sex tourism give the east–west sexual encounter a politically strategic coherence, it is also the case that the prejudicial nature of political activism can alienate activists from their so-called target groups. The anti-military bases campaign was not supported by women in the sex industry, for example, because the political discourse of colonial domination bore little resemblance to the specific and situated way these relations were negotiated in everyday life. Indeed, the families of Alumnos were distressed at the departure of the American military bases as women were forced to return to Cebu to find work in a less vibrant trade.

A cogent feminist-nationalist movement has initiated debates about prostitution, condemning militarism and neo-colonial development policies for the bulk of the sex industry in the Philippines. In Cebu in particular, tourism development policies have facilitated foreign tourist arrivals, displaced urban poor neighbourhoods and exacerbated socioeconomic inequalities in the city. For many women employment in the growing tourist industry is an attractive alternative to lower paying service and industry jobs, but for poor women with limited education the sex industry is one of few viable options. The political and economic dynamics of tourism in Cebu are entangled with women's employment, and communities such as Alumnos increasingly accept sex work as a lucrative way to participate in Cebu's economic growth. It is difficult to draw on structural accounts of political economy, however, and keep the lives of Cebu's bar women in sight. Anti-colonial politics are crucial to building an autonomous national integrity, but they strategically simplify fixed positions and identities of domination/subordination. By homogenizing sex industry workers as exploited victims of neo-colonial development, they also erase the agency of these women.

Employment in the sex industry is only one facet of the lives of women who work there – it is a job women are employed in, usually for short periods of time, thus they do not simply see themselves as 'prostitutes' or 'sex workers'. Just as *inside* the bar women are also dancers, cultural interpreters, prospective wives, etc., *outside* the bar, women are also mothers, daughters, wives, girlfriends, income earners, and so on. In Chapter 4 I explore how these shifting identities intersect women's lives, as well as how these intersections are linked to spaces such as the bar, the neighbourhood, the home, past residences, tourist sites and facilities, foreign countries, and so on. This perspective allows women to be involved inside, outside and *beyond* representations of a victim identity, and to be recognized as constructing the worlds they live in.

By strategically using women's narratives of their entry and employment in the sex industry, Chapter 4 also illustrates how bar women locate their employment as a 'social problem'. These stories give them a space to think, feel and act, and emphasize the richness, fluidity and paradoxes of women's subjectivity. The ethnographic subjects of these narratives are women from the Philippines, and their experiences complement emerging perspectives from studies of prostitution in Indonesia and Thailand. What becomes clear is that these women do not author themselves as victims of political economy nor as part of a global sex traffic. This is not to suggest they enjoy their work, but to highlight their difficulties in identifying with those who want to 'help' them; more specifically, their difficulties in identifying with the politically-inspired projects of non-government organizations.

Educating sex workers: Bangkok, Thailand

At 1:00 p.m. in the afternoon, Patpong is not bustling, showy or extravagant. There are no crowds or neon lights, no street vendors selling fake Rolex watches, and no touts advertising a smorgasbord of sex shows. The doors of the fifty-odd bars are locked, concealing the Sexual Wonderland atmosphere that radiates out each night. If it were not for the names hanging above these establishments – Supergirls, Pussy Galore or Screw Boys – Patpong by day appears rather benign. As I wander around Bangkok's famous red light district, Cleo Odzer's maps in hand,[2] I take note of changes since the late 1980s. The three parallel roads Patpong I, II and III remain intact, although half the clubs have changed names. Odzer's maps do not include the Japanese district, Thaniya Plaza, or the cluster of male go-go bars just east of Suriwong Road. These are the expanding districts of the 1990s as the sex industry captures new markets.

On this particular day I am wandering around in the afternoon before my meeting at the Empower office. Empower is a non-government organization with a drop-in centre for sex industry workers in Patpong. I follow the instructions given to me over the phone, and find their office in the building above King's Corner bar. According to their annual reports, which I have come here to read, Empower was formed in the mid-1980s to address the problems of women in the entertainment industry.

After watching a video about Empower's basic rationale, I read more about their projects and programmes. AIDS education activities began in 1987, after a handful of cases were found in the Kingdom. While Empower offers one-on-one counselling, they have also developed a mobile education unit that uses drama to disseminate information about AIDS. The skits performed portray various characters of the bar scene, and Empower claims this activity alone reaches about 6,000 sex industry workers, 300–400 owners and managers, 150–200 entertainment places and 3,000 customers.

After reading through Empower's reports and some articles from the *Bangkok Post*, a Canadian woman introduces herself as a volunteer English teacher. Empower offers language training in English and Japanese to help bar workers better negotiate money and condom use with their customers, and I am reading in the room where this lesson is usually taught. She suggests I participate since it would be good for the students, and I am grateful for the opportunity.

There are eight of us in total and we are learning very basic question and answer phrases that would be useful in the bar.

Where are you from?
How long have you been in Bangkok?
Are you single?
Do you love me?

Some of the questions solicit hilarious responses, revealing the unambiguous reason for our presence. 'What is your favourite colour?' 'Silver and gold!', 'What is your favourite food?' 'KFC!', 'Are you single?' 'No!'. In the course of the questions, it becomes obvious that the women around me have not been in Bangkok long and have primarily come from Northeast Thailand. They are in their early twenties and alone in the big city, working to earn money to send back home. A few days later I hear a teacher complain about the demand for purely 'bar talk' in Empower's Basic English classes, but these women are not necessarily here for self-improvement or to read AIDS posters on the walls. Many simply want enough phrases to increase their income at work.

Intervening in the lives of women in the sex industry is not a new phenomenon, but Empower's work represents some of the new types of 'education' projects that have emerged in Southeast Asia. Rather than conduct projects that evolve out of anti-prostitution campaigns – such as consciousness-raising seminars or job skills retraining – Empower works within a framework that considers women in the entertainment industry capable of protecting themselves from HIV infection.

Thai government policies encouraged servicemen to engage in 'rest and recreation' activities in Thailand during the Vietnam War, but the sex industry has taken on new and diverse forms with the arrival of mass tourism. Much like the Philippines, political and social activists depict sex tourism as a product of militarism and the political economy of tourism. Sex workers are cast as victims of these structures, as well as overall development strategies that neglect Northeast Thailand. In the effort to slow the spread of HIV, however, and with the influence of international organizations, the types of intervention programmes for these women and the language used to describe prostitution are changing. Because 'empowering' women to protect themselves from HIV/AIDS is considered an urgent priority, perspectives that characterize women as victims have been dislocated by agency-centred, participatory approaches to education. While these organizations and the ethos of their projects have challenged the stereotype of prostitutes as victims, these debates have also tended to reinscribe the dualism of victim/agent, usually around the issue of 'choice'.

In Chapter 6 I outline debates surrounding issues of agency and choice, arguing that organizations such as Empower open up new spaces where it is possible to imagine a different kind of sex worker subject. In Thailand and the Philippines, non-government organizations (NGOs) have been renegotiating the meaning of sex tourism, and some organizations are redefining their community development activities in terms of Western models of education which imagine empowered 'sex worker' subjects. Although these perspectives might eschew discussions of the geopolitical and economic realities which see disproportionate numbers of women enter the sex industry, they have also challenged the frameworks which characterize women in the sex industry as (always) victims of colonialism, militarism and – most recently – globalization. They also address issues in the everyday life of women in the sex industry in ways that can improve their sexual health and working conditions. I use these examples to destabilize representations of sex tourism which marginalize the everyday experience of commercial sex, and to counter representations of AIDS which increase the stigma associated with sex work.

Gendered subjects: Bali, Indonesia

> As an unaccompanied woman on the streets of Kuta, it is impossible to wander alone. In addition to determined traders selling sarongs, baseball caps and Australian cigarettes or newspapers, there are men – commonly known as 'guides' – who proffer their services to see Bali's famed tourist sites. Guides congregate along the main streets of Kuta, usually beside their motorcycles, and offer transport to temples, volcanoes and traditional dance/music. For the single,

woman traveller a trip around Bali on the back of a motorbike can sound like a wonderful idea. What they offer is companionship, cultural interpretation and, in some cases, sex.

Down on the beach there is a slightly different scene on offer. 'Beach boys', as they are known, are surfing Kuta waves while practising English with Australian, European and Japanese women. As they wade into the water or wait for a wave, conversations are struck up about what each is doing that evening. Chances are they will meet up at The Bounty for dinner, head to the Sari Club for drinks and dancing, and end up in the woman's hotel room. While money may not necessarily change hands, beach boys might enjoy a bed to sleep in, meals, drinks and a trip around the island at the woman's expense. If he is lucky, however, she will leave a 'gift' of cash upon departing and include a little money in her letters from home.

Guides and beach boys are two incarnations of the infamous Kuta Cowboy. As I interview Bernie, who is 28 years old, I hear stories of romances with Australian and European women. Bernie grew up in Ulluwatu with his grandmother, since his father died and his mother left him with his paternal grandmother before he was two years old. Ulluwatu is a famous surf beach in Bali, and as a result Bernie learned how to surf. Through surfing Bernie came to meet tourists from around the world, and eventually became a beach boy on Kuta beach. He recalls those days with bewildered amusement, as if it were another version of himself.

Bernie is now married to a Japanese woman, and they have a child who is 1 year old. He gave up his beach boy days almost three years ago, when his wife moved to Bali, converted to Hinduism, and they married. Although he had primarily romanced Australian women during his beach boy days, Bernie claims that Australian women, in addition to being rowdy, tend to dominate Balinese men. Japanese women, on the other hand, are happier in a more submissive role and integrate better because they are culturally similar. His wife is happy to make daily Hindu offerings, have their child brought up as a Hindu, and is adept at making Balinese food. This, he claims, is not possible for Australian women, who are too Western and would not enjoy it.

The problem with marrying a Japanese woman in Bali, Bernie claims, is that all Japanese women are considered to be rich. Everyone in the neighbourhood presumes he's wealthy, even though he just runs a small business and his wife doesn't work. The business he runs sells sarongs for export, particularly to Australia. Bernie designs the sarong motifs, which are mostly of dolphins, but he also has an Australian Aboriginal turtle design he's developed as well.

In the tourist paradise of Bali, sexual liaisons with local men can form an integral part of women's travels to Indonesia. Although less institutionalized than the go-go bar scenes of Thailand and the Philippines, sexual relations between local men and foreign women are touted in tourist guides such as the *Lonely Planet*, and mixed marriages and relationships have become a common feature of Balinese life. Public opinions about tourist relations with Kuta Cowboys have been less political than about other forms of sex tourism in Southeast Asia, partly because this east–west relationship does not coincide with classic representations of exploitation. Nevertheless, when Indonesia's first reported AIDS case was detected in a Dutch tourist in Bali in 1987, Kuta Cowboys became discernible in discussions about HIV/AIDS. Popular associations between HIV/AIDS and foreign tourists were bolstered by HIV cases detected in Indonesia's premier tourist attraction, and government and non-government organizations became interested in education programmes for vulnerable groups such as these Indonesian men. The Citra Usadha Foundation, a small community organization, has initiated 'peer education' activities for beach boys and guides and has four drop-in centres around the island.

The phenomenon of 'romance tourism' in Bali helps to further specify the sex worker subject of AIDS prevention policies. Indeed, the bodies of sex workers are always inscribed with race, gender and sexuality, and these inscriptions help mark them as targets for HIV/AIDS education. In Chapter 1 I examine how these bodies have become objects of institutional power since the advent of AIDS, and what this means in terms of inhabiting bodies which are coded as dirty and diseased. In this way, I ask questions relevant to political discourses/health interventions: what kinds of bodies are imagined, what kinds of men and women, what kinds of prostitute bodies? There are investments in who can speak, who is allowed to speak, and it is important to examine how these boundaries are created and maintained. For representations of sex worker bodies impinge on the lives of men and women engaged in commercial sex, where they are incorporated, contested and/or denied. By focusing on the prostitute body as an object of power, as a site of resistance, as well as a site from which identity is constructed and performed, Chapter 1 provides an alternative approach to understanding the relationship between sex workers and HIV/AIDS.

Relations between Kuta Cowboys and foreign women also highlight how gender and ethnic identities mediate understandings of the foreign sexual encounter. In Bali, for example, Western wealth always mediates assumptions about male dominance – thereby subverting fixed subject positions of powerful male/powerless female. In Chapter 3 I examine how sex tourism districts in Southeast Asia are 'liminal' or 'border' zones, where these sexual, racial and gender differences are worked out. It is here that tourists freely mingle with 'the locals', but where images of sex tourism also mediate these encounters. Although similar cases could be made for Thailand or Indonesia,

Chapter 3 contextualizes this liminality in a go-go bar in the Philippines and describes the bar as a negotiated space of power and identity. I suggests that the white, male gaze is not merely an autonomous voyeurism – it has its own difficulties and uneasiness due to the gaze from supposedly powerless bar women who have their own sites/sights of power, meaning and identity. The bar is one place where stereotypical identities are performed – a 'third space' where Asian bar girl and sex tourist identities are continuously performed, but their fragility can result in the subversion of the 'bad girl', 'exploitative foreigner' or 'white saviour' myths. Examining identity in this way suggests that although images of dominance mediate east–west relations, it is also possible to imagine the politics of the bar outside representations that portray sex tourism districts as sites of foreign oppression.

Global/local sex industries

Sex tourism and AIDS are increasingly generalized as 'global problems', and in Chapter 2 I examine how epidemiological discourse has helped designate Asia as the new epicentre of HIV/AIDS. While mapping might convey an abstract project of locating sex tourism, with the advent of AIDS it is much more than a rhetorical gesture. Widespread concern over identifying and regulating the bodies and spaces of commercial sex suggests that a dynamic understanding of mapping might enable sex industries to be located as new sites of transnational negotiation. With this in mind, I introduce an artefact of the 1990s: a map of the sex industry in Cebu City. This map, and others like it, have appeared on the walls of the offices of government and non-government organizations with substantial sex industries in Asia, sometimes complete with coloured thumbtacks marking sites of AIDS education projects. These maps, and the practices of AIDS prevention they reflect and encourage, have been important in producing a politics of prostitution in the cities they seem to represent. Indeed, the cultural politics of AIDS in Southeast Asia brings together groups as diverse as the state, NGOs, international aid agencies, and groups representing gay or prostitutes' rights movements.

The arrival of HIV/AIDS in Southeast Asia has caused sex industries to come under close scrutiny, and in some cases sex workers have been required to undergo mandatory blood screening for HIV/AIDS. Chapter 5 describes the specific case of the Philippines, and how testing sex workers for HIV fuelled public perceptions of blame. Although prostitution is illegal, the government has monitored sexually transmitted diseases within the 'hospitality business' since at least the 1930s, and women have been required to attend government health clinics on a mandatory basis. Such attendance provided the opportunity to conduct HIV antibody testing in the earliest stages of the epidemic, although there were heated debates about whether or not this was an infringement on women's human rights.

The groups involved in debates on mandatory testing (government, NGOs, the media) constructed prostitution as an object of their discourse, but the debates themselves were not actually *about* prostitution. Instead, mandatory blood testing was about the autonomy of local health officials, and the imposition of Western AIDS prevention agendas in culturally different contexts. This is not to say that these debates were isolated from the lives of women in the sex industry – although they were not included as a voice in the debate – because they played a role in determining national AIDS prevention policies. It is the rather different assertion that debates about the sex industry and AIDS are always saturated with moralisms and political agendas, and sex workers are often used as pawns in the debate.

Sex worker subjects

A major problem encountered in writing about sex tourism in Southeast Asia is describing the phenomenon as a part of a global economy characterized by unequal structures of power and wealth, while at the same time emphasizing the importance of subjective life in empowering people to protect themselves from HIV. *Sex Work in Southeast Asia* examines this problem by outlining different ways of understanding both sex tourism and AIDS as global problems; in other words, to outline the distinct 'maps of meaning' of bureaucrats, activists and women in the sex industry.[3] These perspectives are rarely in conversation, but they combine to mediate the practices of sex work. By bringing these perspectives into dialogue, however, I examine how local and global politics have combined to sustain coercive stereotypes about the sex tourism encounter, and increase the stigma associated with sex work.

One of the aims of this book is to move beyond portrayals of sex tourism and AIDS which use dualistic economies of meaning and power such as global/local, west/east and, often by implication, powerful/powerless. These portrayals neglect the worldviews and everyday experiences of Southeast Asian sex workers, who rarely consider themselves victims of the political economy or part of a global sex traffic. Instead their lives are framed by issues of employment opportunities, family responsibilities and dreams of a better life – at home or abroad. Furthermore, their relations with foreign tourists are often understood in romantic or benign terms, where paid sexual encounters are enmeshed with their exit from the industry. These perspectives play an important role in the practice of commercial sex, although AIDS prevention programmes that encourage women to leave the industry or to negotiate condom use for each sexual encounter often neglect them. They are also neglected by activist discourse that mobilizes images of sex workers as victims to politicize the issue of commercial sex.

The book's intentional appeal to a spatialized understanding of these debates enables sex work to be understood as a site of negotiation where local, national and international debates intersect in new and unfamiliar ways. Furthermore, notions of community and scale can be introduced in ways that also account for the organization of the book – that is, from global mappings of red light districts to the contexts of individual bars, and from neighbourhood conversations to national debates. In emphasizing how the cultural politics of AIDS in Southeast Asia selectively incorporates initiatives from a transnational arena, I suggest that sex industries must be mapped within a complex grid of multiple, intersecting and translocal relationships where it is possible to imagine less determined spaces of subjectivity for Southeast Asian sex workers. It is only in these spaces, I argue, that the voices of sex workers can be heard.

Mediating the tensions between the subjective lives of sex workers and the instrumental discourse of government and non-government organizations holds the possibility of finding new sites of collaboration and contestation in contemporary debates on prostitution, with new perspectives on identity, power and resistance. The concluding chapter examines the question: if the dominant construction of prostitution does not adequately reflect the lived experience of sex work, what alternatives exist? I argue that the 'peer education' model adopted by some Southeast Asian NGOs has the potential to hold these perspectives in tension, but that these projects occur in an international context of rational decision-making that is often at odds with subjective life. Yet by exploring the tensions and ambiguities of these subjectivities, this book offers an alternative way to imagine sex work in theory and practice.

1

RETHINKING THE PROSTITUTE SUBJECT

Bodies, subjectivities and space

> By body I understand a concrete, material, animate organisation of flesh, organs, nerves, muscles, and skeletal structure which are given a unity, cohesiveness, and organisation only through their psychical and social inscription as the surface and raw materials of an integrated and cohesive totality . . . The body becomes a human body, a body which coincides with the 'shape' and space of a psyche, a body whose epidermic surface bounds a psychical unity, a body which thereby defines the limits of experience and subjectivity, in psychoanalytic terms through the intervention of the . . . Other or Symbolic order (language and rule-governed social order).
>
> (Grosz in Longhurst 1995: 98)

While there remains little consensus on the question 'what is the body?', the subject of the body has captured the attention of social theorists for the past decade. In concert with critiques of Enlightenment rationality, many argue that the belief in a disembodied, rational knowledge silenced – or at least severely muted – the role of the body in social life. The body thus became the container for consciousness, the impediment to reason and, ultimately, the site to be transcended. This mind/body split encouraged the conviction that there was only one 'pure' or 'perspectiveless' kind of knowledge – that there was only one Truth.[1]

The gendered character of the mind/body dualism has inspired some feminist writers to locate the body at the centre of analysis. Departing from notions of universal experience and truth, these subject-centred theories emphasize the fluid construction of personal identity and accentuate the ways individuals make sense of the social world from unique and variegated 'subject positions'. Rather than depict the mind as the irreducible unit of subjectivity, the body is recognized as a crucial site where power and knowledge are negotiated. The experience of subjectivity is therefore bound up in orders of power where bodies are moulded and policed by social, psychic and cultural norms, and where the power/knowledge nexus interpolates the subject.

Recognizing the existence of multiply constructed realities, and identifying the power/knowledge regimes that allow particular views to gain currency, is important in the context of power, sex work and HIV/AIDS. On the one hand, there is a need for a language of power and position, since sex workers have become the targets of local, national and global HIV/AIDS education campaigns. The normativities of HIV prevention assume sex workers require education, thereby ascribing a stigmatized sex worker identity: the female, Southeast Asian prostitute is understood as infected and infecting. On the other hand, these regimes of power and knowledge have resonance in everyday life. They police the practices of commercial sex, for example, as well as the experience of inhabiting a body coded as diseased.

Thinking through the specificities of bodies and spaces provides a means to imagine a sex worker subject which is simultaneously located in rigid grids of power, meaning and identity (e.g. male/female, Western/Asian, healthy/diseased), but is also fluid, multiple, and capable of change and mobility. By focusing on the prostitute body as an object of power, a site of resistance, as well as a site where identity is fluidly constructed, this chapter provides a less deterministic approach to understanding the relationship between sex workers and HIV/AIDS.

The body and Foucault

> I wonder whether, before one poses the question of ideology, it wouldn't be more materialist to study first the question of the body and the effects of power on it. Because what troubles me with these analyses which prioritise ideology is that there is always presupposed a human subject on the lines of the model provided by classical philosophy, endowed with a consciousness which power is then thought to seize on.
>
> (Foucault 1980: 58)

Understanding the human body as both an object of power and the site at which power is disseminated and resisted owes much to Foucault's work on the normalization of bodies in prisons, schools and hospitals. Foucault's work on regimes of disciplinary control – notably his writing on criminality, sexuality and mental illness – highlights how the sociopolitical structures which organize the practices of everyday social life produce knowledge about particular bodies and influence the ways bodies experience time and space. For Foucault, the body is a 'surface of inscription' marked by dominant cultural norms (of criminality, sanity or sexuality), but holds the possibility of inscribing itself on social practices through resistance. Instead of emphasizing how bodies resist and subvert power, however, Foucault's emphasis is on how particular forms of power can manipulate the 'political economy of the body' to produce bodies which suffer systemic subordination.

In *The History of sexuality*, for example, Foucault (1978) examines the relationship between the body and sexuality, critiquing the 'repressive hypothesis' of the Victorian era and arguing for a different conception of the relationship between changing sexual norms and the power of discourse. He differentiates between sexual practices (which are performed by the body) and sexuality (which is discursively constructed), arguing that it was the Victorian era which named diverse sexual practices, assigning them values and meanings, so that particular bodies were observed to contain particular socio-sexual identities. It was in this discourse of *scientia sexualis*, that powers of surveillance and regulation produced knowledge about particular kinds of bodies. Similarly, in *Discipline and Punish*, Foucault (1977) argues that the disappearance of bodily torture and its replacement with institutions of discipline was correlated to the state's need to control individual bodies (e.g. at work and in schools, the military, hospitals, etc.). The institutionalization of the power to punish – in this case, the rise of the French penal system – was more effective in producing knowledge about a pathologized subject ('criminals'), however, than it was in reducing the crime rate. Bodies in the prison were enclosed, partitioned and ranked, and their activities were controlled in time (timetables) and in space (the panopticon) to produce what Foucault terms 'docile bodies'. This meticulous control produced bodies that served as the intermediaries of punishment; that is, the prisoner's body was disciplined in order to correct their behaviour (or what Foucault refers to as the 'modern soul').

It is tempting to use this approach to understand the context of prostitution and HIV/AIDS in Southeast Asia. A pathologized sex worker subject is clearly discernible within a series of institutional discourses that seek to identify, name and regulate the sexual practices of these bodies. There are libraries of demographic and social information describing the places and people of sex work, and groups ranging from the World Health Organisation to academics to community-based organizations have been involved in data collection and surveillance. In naming the sexual practices that sex workers engage in with their clients, as well as testing their blood for HIV antibodies, the discourse of AIDS prevention constructs a socio-sexual sex worker identity and produces knowledge about prostitution in Southeast Asia.

A Foucauldian analysis of sex work relies on a very specific conception of power, however, one which has come under increasing scrutiny (Hartsock 1990). Sex workers merely resist the power of institutional discourse, rather than pursue different ways to redefine their social world. Furthermore, Foucault conceptualised power–knowledge–pleasure – and more specifically, desire – within a framework that neglected how gender and race were constitutive of particular workings of power (Stoler 1995). Because sex tourism and HIV/AIDS prevention are issues which are simultaneously about gender and race, a purely Foucauldian analysis neglects the place of desire in imagining sex worker bodies.

The body and feminism

> As soon as knowledge is seen as purely conceptual, its relation to bodies, the corporeality of both knowers and texts, and the ways these materialities interact, must become obscure.
>
> (Grosz 1994: 4)

The issues of power/knowledge, exclusion and invisibility have been theorized by Western feminists for several decades. Broadly speaking, these feminists have sought to identify and redress the inequalities of a patriarchal culture that denies women a meaningful role in the production of social, economic and cultural life. While homogenizing Woman as a universal category was a strategic (and contingent) move with its own imperatives, prevailing modes of representation and knowledge formation tended to reproduce this Woman within the realm of masculinist discourse. As a result, the imperatives of only *some* women were theorized and represented under the assumption of a dominant reality of female oppression. Resistance to such exclusion became evident through critiques from women of colour, lesbians and women in the sex industry, and the ethnocentric, heterosexist and middle-class character of Western feminist analyses was revealed. In response to these critiques, and by accepting the fiction of unity, some feminists redirected attention to how 'gender' is multiply intersected by race, class and sexuality. This is a positive and productive move that encourages a re-imagining of Woman, but the possibility of relativism is always present: what are the political risks of deconstructing Woman? Does multiplicity become a chaotic proliferation of voices that produce a noisy babble?[2] These questions are in concert with the current unease over the fate of 'the subject'.

Feminists of difference – that is, feminists inspired by psychoanalysis, Derridean deconstruction and Foucauldian genealogy – have taken a less pragmatic approach to the issue of female oppression, producing complex philosophical readings of Woman within a highly spatialized language of bodies, subjectivity and sexual difference.[3] Feminists of difference conceive the female object/male subject dichotomy in terms of a psychoanalytic understanding of 'othering'; that is, where 'othering' is recognized as a mapping of self. Within this framework Woman is defined in terms of her negativity (as not-Man) in the dominant Symbolic order and, in a more spatial sense, forms the boundary to what constitutes masculine identity. In this way, Man cannot exist without (the exclusion of) Woman. Instead of focusing on how women appropriate, resist or subvert this negatively defined subject position, feminists of difference seek to transgress the dualism though imagining what might *exceed* these analytic processes. 'Bodies' – or corporeal subjectivities – constitute this excess, and are important in imagining what a more autonomous feminine subjectivity might look like. This is a contested move which has been criticized as essentialist, but if understood as situated and contingent, holds the possibility of imagining a space for a female subject outside masculinist discourse.

Bordo (1992) is correct in asserting that it was not Foucault who 'discovered' the politics of the body, but it was during the late 1980s and early 1990s that feminists began conscientiously experimenting with the concept of a 'corporeal feminism' (Gatens 1988; Gallop 1988; Bordo 1989; Grosz 1989, 1994). Feminists had been aware of the bodily nature of women's disempowerment – exemplified in protests against pornography and calls for the self-control of reproductive functions (e.g. birth control, abortion) – but had been reluctant to theorize *from* the body, as an embodied subjectivity, for fear that it would naturalize Woman's social position. Much discussion in the 1970s and 1980s was therefore devoted to the social construction of gender. It was not until the critique of the sex/gender split emerged (Gatens 1988; Haraway 1991; Butler 1990, 1993), however, that sexed bodies were seriously considered as starting points for feminist analysis.

Broadening Foucault's conception of the dialectical relationship between power and knowledge – and the role of the body within that relationship – some contemporary feminists seek to theorize the role of the *sexed* body in the production and evaluation of knowledge(s). Indeed, Foucault has been criticized for neglecting precisely this issue (Probyn 1991; Grosz 1993). This means understanding the construction of the female body not only as the site of the 'other' – that is, the site of difference categorized as irrational and emotional in the quest to equate masculinity with rationality and vice versa – but more generally as a politically inscribed and discursively constructed object of power, and as a site of discipline and disempowerment. This emphasis denaturalizes women's bodies and social position by focusing on the technologies of power which 'create' women as 'subjects' in the first instance.

This imagining is taken further by Irigaray (1985) and Grosz (1994), who argue for a 'corporeal feminism' which uses the female body and an embodied, non-dualist subjectivity as starting points for analysis. This approach reduces the likelihood of the feminine being reproduced within the realm of masculinist discourse, particularly the reproduction of the male/female dualism where Woman is considered a knowable object, and Man the knowing subject. Instead, it opens the space for multiple corporeal subjectivities. This has serious implications for feminist thought, however, because it means dispensing with the universalizing category of Woman. Deconstructing Woman further problematizes the position of the knower, and in so doing critiques particular kinds of knowledge formation. Unlike Foucault, therefore, feminists offer a more dialectical and transformative understanding of the body. According to Grosz (1993: 187–8),

> [if] the body is always sexually specific, concretely 'sexed', this implies that the hegemony over knowledges that masculinity has thus far accomplished can be subverted, upset, or transformed through women's assertion of 'a right to know', independent of and

> autonomous from the methods and presumptions regulating the prevailing (patriarchal) forms of knowledge.

There have been various interpretations of the placement of the sexed female body at the centre of political and theoretical debates (Grosz 1994: 14–19). These interpretations, Grosz argues, can be broadly categorized as emanating from three sources: egalitarian feminism, social constructionism and those advocating a conception of sexual difference. Egalitarian feminists[4] perceive the female body and its attendant reproductive functions as simultaneously a hindrance to equality between the sexes and an opportunity for women to gain special insights to life. This perspective therefore perceives a conflict between woman as mother and woman as political being (i.e. private and public roles), necessitating reproductive technologies that enable women to participate more fully in the public sphere. As Grosz points out, however, this perspective adheres to the notion that the body must be transcended, maintains the public/private dichotomy, and does not question why maternal or potentially maternal bodies are excluded from the public sphere in the first place.

Social constructionist feminists,[5] on the other hand, have focused more on the gendering of productive (i.e. male) and reproductive (i.e. female) roles, aiming to minimize sexual differences and reassign new and more positive values to the latter (e.g. child rearing). While this is a more positive attitude towards the female body, it presumes that sex is a fixed category and gender is the area of transformation. This sex/gender opposition adheres to the body/mind dualism, however, where 'real' bodies and the representations of them are mutually exclusive. Finally, to feminists who adhere to the notion of sexual difference,[6] understanding the *lived body* is central not only to transgressing this mind/body split, but to contemporary (political, sexual and intellectual) struggles. These feminists argue that the body is the object of power, but also the site of dualisms themselves, and are therefore more concerned with the body as a political, social and cultural – as well as discursively constructed – object: 'a body bound up in the order of desire, signification, and power' (Grosz 1994: 19).

I am uneasy with such a rigid categorization of feminists, since there is more epistemological similarity between these perspectives than I account for here (cf. Fuss 1989; Rose 1993a). But there is an important space opened up by feminists adhering to the notion of sexual difference. Within this perspective, there is a possibility of multiple types of bodies, and hence types of subjectivity, thereby allowing the body to be the 'mediator' of experience. By conceiving of the body as 'a point of mediation between what is perceived as purely internal and accessible only to the subject and what is external and publicly observable' – that is, by problematizing experience as access to Truth – this allows a remapping of other dualisms such as inside/outside, public/private, self/other, and a host of other binary pairs associated with the mind/body split (Grosz 1994: 20–1). In Southeast Asia, understanding the lived body allows me to conceptualize sex workers without separating out categories of class, race and gender as if they were mutually exclusive. It also allows me to interrogate and transgress the dualisms implicit within current debates on sex tourism and HIV/AIDS, such as rich/poor, west/east, agent/victim, healthy/diseased, and so on.

Feminists who advocate an embodied subjectivity as a starting point for analysis have been criticized as essentialist, for conceiving of the body too excessively as a text, and for being too concerned with rewriting the body as a positivity rather than a lack (Bordo 1992; Rose 1993a; Mortimer 1994/5). As Grosz (1993: 195) points out, however, accounts of bodies are not *only* anatomical, physiological, or biological, they can also be *sociocultural*: 'nonbiologistic, non-reductive accounts of the body may entail quite different consequences and serve to reposition women's relations to the production of knowledges.' Furthermore, if power is seen to operate directly on bodies, then understanding the relationship between bodies, power and knowledge production acknowledges the (negative, subordinated or excluded) term *body* as the *unacknowledged condition* of the (dominant) term reason (Grosz 1993: 195). Finally, and with regard to rewriting the body as positivity rather than lack, at issue is the question of agency. Grosz (1994: viii) has suggested that 'the notion of agency . . . can be remapped, refigured, in terms of models and paradigms which conceive of subjectivity in terms of the primacy of corporeality, which regard subjectivity on the model not of latency or depth but of surface.' In terms of concrete political action, however, there is concern that this approach does not emphasize the structures that impede women's agency. Indeed, and particularly in the case of women employed in the sex industry, there is a need to devise frameworks for understanding women's autonomy and agency which transgress the structure/agency dichotomy.

There is overlap between Foucault's 'intermediary' bodies of discipline, and Grosz's body as 'mediator' of experience; indeed, they are the sites at which the performance of power, knowledge and resistance are worked out. Their emphasis differs, however, in that Foucault is interested in the classical age's discovery of the body as object and target of power, while Grosz is more concerned with elaborating a corporeal subjectivity to critique particular kinds of knowledge production. Both concepts can be mutually reinforcing, however, and 'the body' thus provides a useful starting point for understanding the relationship between power, knowledge and prostitute bodies.

The body and prostitution

In an extensive scholarly work, Bell (1994) considers the subject of 'the body' in Western conceptualizations of prostitution at different historical junctures. Moving through the discursive construction of the prostitute in Plato, modernism, feminism and by international prostitute organizations and performance artists themselves, Bell argues that a process of 'othering' is evident in modern and feminist constructions of the prostitute body – dichotomizing women into good/bad, healthy/diseased and agent/victim. Prostitute organizations that seek to break such dichotomies are often prone to reproducing them, and Bell (1994: 2) argues that it is only in prostitute performance art 'that the overarching dichotomies of "whore" and "madonna" . . . are dissolved and unified'.

Following Foucault, Bell claims that it was in the modern period that 'the prostitute body was produced as an identity and prostitution as a deviant sexuality' (1994: 12). She analyses the medical–moral–legal discourses of modernity in the popular texts of Parent-Duchatelet, William Acton, Havelock Ellis and Freud, and argues that throughout the modern era:

> the prostitute body was produced as a negative identity by the bourgeois subject, an empty symbol filled from the outside with the debris of the modern body/body politic, a sign to women to sublimate their libidinal body in their reproductive body.
> (1994: 72)

Her careful re-reading of these texts reveals more ambiguity than has been previously offered, but this does not detract from her argument that the systematic surveillance and control of prostitutes in the modern era (by doctors, members of the police and clergy, reformatory matrons, sexologists and psychoanalysts) were an assertion of power over the 'other' that resulted in much knowledge production about prostitutes and prostitution.

Using the texts of Pateman (1988), MacKinnon (1987, 1989), Irigaray (1985) and Rubin (1984), Bell analyses feminist (re)productions of the prostitute body. She argues that the 'sexual contract' theory offered by Pateman (which accepts the dichotomization of women into prostitutes and wives), and MacKinnon's notion that female sexuality is constructed as an object of male desire (thus reproducing the prostitute as women's fundamental position in masculinist society), are both inherently Marxist in their approach, equating the selling of labour power with exploitation. By 'equating sexuality with labour, desire with value, and "women's point of view" with class consciousness', Bell (1994: 82) argues that these feminist writers silence the voice of 'other' women (i.e. prostitutes) and reproduce the prostitute body within the realm of masculinist discourse. Other critiques of feminist analyses of prostitution concur with this view (Bell 1987; Truong 1990; Delacoste and Alexander 1987). Irigaray's conception of the 'sexual female subject' and Rubin's 'democratic morality', on the other hand, open the space for a multiplicity of sexual voices, thus positioning the prostitute as a speaking subject. It is from this perspective that Bell introduces us to prostitutes' rights organizations in the West, and their various theories of prostitution.

Bell (1994: 99) emphasizes that only in postmodernity has the prostitute emerged as a political and plural subject. She demonstrates this thesis through an examination of different types of discourses emerging from prostitutes organizations' in the West, although her analysis is also applicable in new debates on prostitution and HIV/AIDS internationally. She uses three examples: prostitutes' rights groups,[7] WHISPER (Women Hurt in Systems of Prostitution Engaged in Revolt) and Prostitutes Anonymous. Their respective perspectives differ in that:

> Prostitutes' rights groups are concerned about the empowerment of the prostitute in the context of a patriarchal and capitalist society; WHISPER contends that the prostitute can never be empowered until she ceases to be a prostitute. Prostitutes Anonymous is concerned about prostitution insofar as it has become an addictive-type problem for the individual prostitute.
>
> (Bell 1994: 99–100)

In demonstrating how these three perspectives draw from very different sources of theorizing – while at the same time reproducing the 'prostitute as powerful sexual being' versus 'disempowered sexual victim' – Bell emphasizes the heterogeneous and contested nature of current discourses of prostitution. This produces an ambiguous identity for the prostitute, but it is only in prostitute performance art that this ambiguity is developed and held fully.

Prostitute performance art is able to hold the victim/agent dichotomy in tension because it is a spectacle of women talking about their lives as prostitutes. It is in this context that the prostitute is author and subject, while at the same time discussing the difficulties of working in prostitution on her own terms. Yet it is also an opportunity to present this body not merely as its representations in modernist and feminist discourse – for example, as obscene or as 'the other' – but as an artist, an erotic being, an intellectual and social commentator (Bell 1994: 142). This representation successfully challenges abstract categorization, and performance artists use their bodies as sites of resistance.

Bell's work is pioneering in many respects, and her conception of a speaking prostitute subject provides an innovative way beyond dualistic thinking on prostitution. It is only when she discusses prostitute organizations and performance artists, however, that Bell gives us any idea what an embodied prostitute subjectivity might be. The crucial point that the body is the site of the dichotomies themselves is treated only in a limited manner, and is restricted to a discussion of theatre/spectacle. How these dualisms might affect the performance of prostitute identity in different spaces is left untheorized, perhaps because of the exploratory nature of the text. I aim to redress this gap through drawing on the literature on the body in geography, where some original work on the spatial nature of identities and places is ongoing.

The body and geography

Geographic thought has been much influenced by recent developments in social theory. In particular, geographers have argued that the postmodern emphasis on the situated and context-dependent nature of social activity necessarily marks an era that concedes the importance of space and place, and therefore geography. This theorizing has occurred within the context of what might be called post-Marxist geography (Harvey 1989), the history of geographic thought (Gregory 1994) and the geography of urban environments (Soja 1989). But there has also been much theorizing within the historically more marginalized areas of geography, such as in feminist, cultural and gay and lesbian studies (see, for example, Watson and Gibson 1995; Bell and Valentine 1995). It is within these margins that the discussion on 'the body and geography' is taking place. These geographers are interested in understanding the intersection of the debates on the body, sexuality, identity and space/place.

Feminist geographers have an ongoing interest in broadening feminist scholarship within the discipline by deconstructing what are perceived to be hegemonic, masculinist and disembodied geographical knowledges (e.g. Massey 1994; Longhurst 1994, 1995; Rose 1993a, 1993b; Johnson 1994; Valentine 1993). They are keen to reject the assumed 'transparent space' described by humanistic geography, which is produced from 'assuming a vantage point far removed from the embodied social world' (Rose 1993a: 71). Such theorizing has led these geographers to consider the sexed body as a theme in their empirical work. Longhurst (1994; 1995) and Rose (1993a), for example, interrogate the body in their work on maternal bodies in space. Longhurst (1994) in particular provides an interesting account of how the (de-centred) body subjectivity of pregnant women affects their movement through the city. She interrogates how certain spaces either reject or welcome these women – due to built form or to acceptable behaviours within particular spaces – and suggests further work should consider how space also constructs (pregnant) bodies with particular desires and capacities. From a feminist/lesbian perspective, but continuing the theme of bodies, identity and place, Valentine (1993) analyses lesbian perceptions and experiences of space. She argues that the norm of heterosexuality is powerfully expressed in everyday spaces – such as neighbourhoods, the workplace, service and commercial environments and public open spaces – often limiting the spaces that lesbians feel comfortable in, and therefore inhabit.

Writing on the cultural politics of masculinity, Jackson (1991) introduces the concept of a plurality of masculinities that influences the way different men move through space. He argues that although those interested in sexual politics have begun to ask 'new questions . . . concerning the ways in which masculinity is embodied, how male subjectivities are constructed and experienced' (Jackson 1991: 210), there is a need to explore the instabilities and contradictions inherent in the notion of masculinity itself. In a similar vein, but drawing more on the notion of gender performance outlined by Butler (1990), Bell *et al.* (1994) examine the identities of the hyper-masculine 'gay skinhead' and the hyperfeminine 'lipstick lesbian' in their relation to space. More specifically, the authors question whether the parodic nature of these identities actually challenges the notion of 'straight (heterosexual) space'. While the authors are interested in examining the relationship between gendered spaces and the spatial nature of gender identity, they highlight the 'performative character of both gender and of place' (Bell *et al.* 1994: 33) in that heterosexual performances create heterosexual spaces and vice versa.

Each of these geographical studies is admittedly different, but I have outlined them here to emphasize their shared commitment to understanding the sexed body in space. For it is bodies – not essentialized bodies but bodies with divergent and multiple identities – that move through landscapes of power and knowledge. Men, women, gays, lesbians, pregnant women, and so on move through these landscapes, and not only do their mobile (and constrained) bodies create space through the performance of gestures that relate to identity, these spaces in turn construct bodies with particular needs and desires. The power of discourse and its relation to the body are therefore enmeshed with space, place and geography.

Prostitute bodies as corporeal subjects

By acknowledging the importance of corporeal subjectivity, as well as the differently constructed voices of women, it becomes possible to conceptualize bodies as 'maps of power and identity' (Haraway 1990: 222). These body-subjects are located in multiply intersecting grids of power, meaning and identity, and it is through what Pile and Thrift (1995: 39–44) term 'encounters with others' that bodies themselves become 'a point of capture where the dense meanings of power are animated'. Encounters tend to animate certain intersections, and create a space where the more immediate boundaries between self and other regulate meaning. Within such encounters, however, there is always excess – a series of differences that play a less visible role in the negotiation of meaning. These excesses can be thought of as a space where it is possible to re-imagine the subject. This is an important point: it enables feminist analyses to be extended to different kinds of social encounters (between and across race, class and gender), and it refuses the notion of an essentialized subject.

The intersection of current debates on power/knowledge, sexuality, identity and space is relevant to prostitution in Southeast Asia. Drawing from the theorizing of Foucault and contemporary feminists, I maintain that prostitute bodies are irreducible: they are the objects of power and sites of resistance, and the sites at which these struggles are worked out. Prostitute bodies are the objects of a disciplinary power that polices morals and public health, and become marked by the practices they engage in (e.g. when and where they engage in business, who their customers are, whether or not they use condoms, etc.). These practices are surveyed, regulated and controlled – usually but not exclusively through the machinery and technologies of the 'state' – and this exertion of power creates much knowledge about a pathologized subject ('prostitution'). The discourse of prostitution therefore produces a singular and unambiguous prostitute identity that is reflective of the (sexual) practices in which they engage.

Yet if identity formation is interpreted as a complex, dialectically constructed relationship between particular forms of power/knowledge and experience, an understanding of this process has the potential of subverting specific kinds of knowledge formation about prostitutes and prostitution (e.g. through AIDS education projects). It is for this reason that I move from discussions initiated by feminists of difference to the literature in geography, where an emphasis on the relationship between identity and space is considered pivotal. Geographers highlight how bodies make particular movements through space; bodies therefore have different experiences and subjectivities. The spaces through which bodies move, however, are considered far from innocent; they are landscapes of power that also mould and police subjectivity. These spaces welcome some and reject others, affirm norms of behaviour and appearance and are inscribed with the dominant cultural norms (of morality, sexuality, and so on). If identity is experienced as movement, and space is reflective of cultural norms, then actual movements through space, and the places created and experienced in this process, work in a dialectical fashion that mediates the experience of identity.

The bodies of Southeast Asian sex workers make movements through space: through red light districts, health clinics, through local communities and the spaces their customers also inhabit (e.g. clubs, hotels and beach resorts). Sometimes these movements correlate directly with the exertion of power. In the Philippines, for example, sex workers are required to attend the City Health Department on a weekly basis, and this space equates identity with sexual practices thereby evoking specific experiences of identity. Other spaces, such as the bar or neighbourhood, evoke different experiences; they can simultaneously reinforce or evade the identity of sex work, depending on the context. It is because these spaces are also situated within other spaces of political, moral and religious significance, however, that they affect experiences of identity. Red light districts are mapped by global AIDS prevention agencies, for example, and the behaviour of sex workers is policed by dominant codes of morality. Sex work is variously considered immoral, sordid, dirty, and so on, and while sex workers themselves might reject these labels, they are nonetheless encountered on a regular basis. Indeed, they constitute the landscapes of power and identity that are interrogated in this book. It is this notion of the construction of identity and its relation to different spaces that helps refuse the presumed (and universal) prostitute identity articulated in debates on prostitution and HIV/AIDS.

2

CARTOGRAPHIES OF DESIRE

Mapping Southeast Asian sex industries

> [A] map does not simply describe what is. A map does not only set up a grid which determines what can be found by selection or omission. Nor is it merely a series of lines inscribed on a previously blank surface. There is an alterity which provokes the desire to map, to contain and to represent – which is to say, to make familiar . . . Mapping as representation is inextricably caught up in the material production of what it represents.
>
> (Diprose and Ferrell 1991: ix)

In the 1990s the epidemiological map of the world marked Asia as the epicentre of HIV/AIDS. This reflected the focused attention of multilaterals such as the World Health Organisation (WHO), as well as governments and community-based organizations seeking to reach marginalized groups with education activities. While the imperatives of these institutions varied from global/national policy to grassroots community education, what they shared was a desire to locate individuals vulnerable to infection. The early detection of HIV cases in Southeast Asian tourist sites, and particularly amongst foreign nationals, played an important role in how the cultural politics of AIDS in Southeast Asia would unfold. Somewhat predictably, AIDS prevention policies and education activities focused on women in the sex industry. Combining international expertise, health department statistics and countless survey findings, sex workers in Southeast Asia were brought into the global discourse of HIV/AIDS prevention.

There is an important cultural critique of AIDS prevention policies that are aimed at particular groups and individuals within societies.[1] These critiques have been instrumental in highlighting how marginalized groups are conceptualized as 'objects' and not 'subjects' of AIDS prevention, and how this regularly leads to blame and stigma. There were many debates about whether sex workers in Southeast Asian tourist sites were culpable, for example, but their designation as a 'high risk group' had more to do with existing anxieties about sex tourism than the sexual health of sex workers. While activists and bureaucrats considered what kinds of education were necessary, the debates themselves exceeded discussions about the sex industry or AIDS. Evaluating the status of the sex industry in the

1990s was simultaneously about the negative consequences of the global links in the region, and the stigma associated with women who engage in sexual relations with men.

While cultural critiques of AIDS have done much to highlight how the designation of 'high risk' groups has been enmeshed in the politics of class, race, gender and sexuality, there has been less attention paid to how cultural politics shape interventions in places outside the West. Furthermore, critiques that do focus on the cultural politics of AIDS in Asia tend to emphasize how the epidemic has been represented in Western texts (Nguyet Erni 1997; although see Buckley 1997 on Japan). Yet the cultural politics of AIDS in Southeast Asia shares the 'othering' evident in Western HIV prevention policies, and is enmeshed with the cultural normativities of global imperatives to slow the spread of HIV. Here I extend these critiques to examine the dynamics through which one group of women in Southeast Asia have been scapegoated for the spread of HIV.

In addressing these issues, I use a dynamic conception of mapping to characterize the various ways of locating both sex tourism and HIV/AIDS as global problems, examining how biomedical and activist discourse helped (re)constitute the politics of prostitution. While mapping might convey an abstract project of locating sex tourism, with the advent of AIDS it is much more than a rhetorical gesture. Widespread concern over identifying and regulating the bodies and spaces of commercial sex suggests that a dynamic understanding of mapping might enable sex industries to be located as sites where local, national and international debates intersect in unfamiliar ways. With this in mind, I introduce an artefact of the 1990s: a map of the sex industry in Cebu City, Philippines (Figure 2.1). This map, and others like it, have appeared on the walls of the offices of government and non-government organizations with substantial sex industries in Southeast Asia, sometimes complete with coloured thumbtacks marking sites of AIDS education projects. These maps – and the practices of AIDS prevention they reflect and encourage – have been instrumental in producing a politics of prostitution in the cities they seem to represent.

In the Philippines, for example, the cultural politics of AIDS echoes an assertion of, and resistance to, imperial ideology and practice. In many ways AIDS education campaigns are reminiscent of earlier attempts to stop the spread of sexually transmitted diseases in colonial times (Manderson and Jolly 1997), and there has been resistance to the deployment of Western expertise and models of education to slow the spread of HIV. Hence it could be argued that the activities of the WHO and related organizations occur within reconfigured power relationships in an era of late capitalism, where public health imperatives are devised by those with the knowledge, power and money to envision and fund projects. In other words, the WHO could be seen as a reified version of the West, where managing the spread of HIV/AIDS is a project of Western desire. But AIDS education policies are no longer

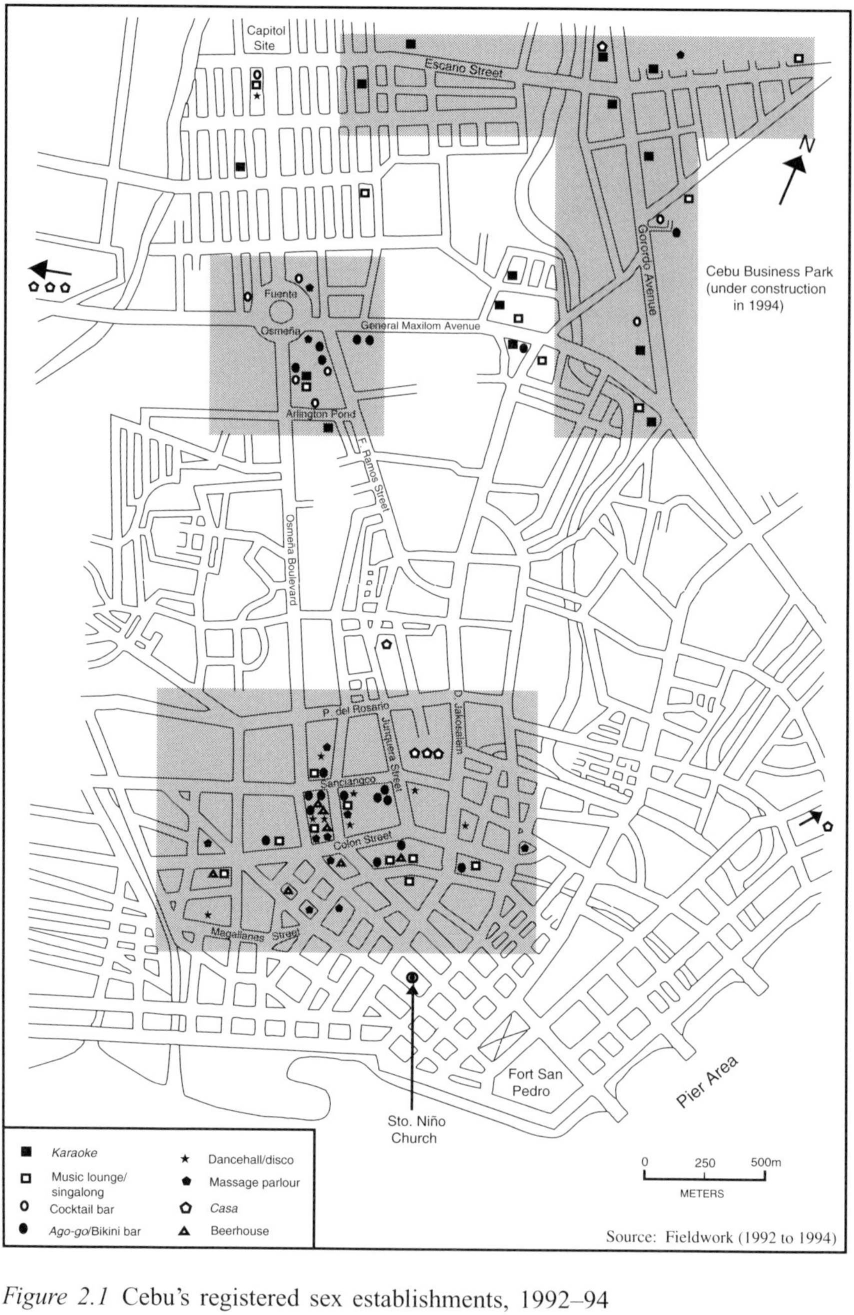

Figure 2.1 Cebu's registered sex establishments, 1992–94

mandated or carried out by colonial authorities, and the ways in which contemporary economic, social and political issues combine make them not clearly recognizable through binaries such as global/local, colonizer/colonized or West/East. Indeed, one of my aims is to unpack these postcolonial subject positions and relations in ways that remain committed to a progressive politics.

These conceptions of culture, politics and space are important since they have helped to define a geography of AIDS which identifies (and blames) particular individuals and groups. Although the focus here is on women in the sex industry, an analogous case might also be made for the designation Asia as the new (i.e. post-African) 'epicentre' of the virus. While cartographic constructions of space have been acknowledged as being integral to the deployment of colonial/imperial power – indeed, the map has become the 'over-determined signifier of the spatiality of the imperial imagination' (Jacobs 1996: 20) – I would also argue that the production of Cebu's sex industry map is important, controversial and ambivalent. Understanding the politics of prostitution in Cebu is not a process of uncovering some mappable reality, nor is it a voyeuristic journey into the world of commercial sex. My aim, following urban-culturalists such as Jacobs (1996: 6), is to 'uncover the unsettled geography of space and place and the shifting boundaries of identity', and to approach these in ways which illustrate how colonial anxieties persist in the contemporary moment – shaping the politics of place and identity in Cebu.

Cebu's sex industry landscape

I have opened this chapter with a quote from Diprose and Ferrell to highlight two interrelated issues: mapping as representation and the material production of space. In terms of representation, I use an interpretive mode to describe what often passes as an abstract and homogenous 'red light district'. I initially understood Cebu's sex industry through the City Health Department, their statistical data and a rather epidemiological focus on what they were learning to define as the 'commercial sex industry'.[2] As I became more familiar with political activists, however, my perspective of the industry altered. No longer did I conceive the sex industry as a set of establishments where employees engaged in risky health practices; instead such establishments were inscriptions of foreign control in the landscape, enabled by government policies on tourism. These perspectives are not necessarily autonomous, but they are divergent discourses that inform debates about prostitution in Cebu. They also demonstrate how knowledges about the industry are 'situated' (Haraway 1988); that is, how the sex industry can be constructed in different ways in different contexts.

In addition to being situated, these particular perspectives are also constitutive of power, and influence the way places such as red light districts are bounded, experienced and understood. Both of these groups, for example, participate in AIDS education activities which eventually intervene in the everyday lives of women in the sex industry: the City Health Department requires women to attend the STD/AIDS Detection Clinic on a weekly basis to be tested for sexually transmitted diseases, and on a biannual basis for HIV antibody testing; and activists frequent nightclub establishments to conduct informal AIDS education activities, as well as visiting the communities where sex workers live. Both groups therefore engage in relations with these women that produce information about 'the sex industry', and the knowledges produced are informed by, and help to constitute, a geography of commercial sex.

Before pursuing these perspectives further, however, I present my own understanding of the sex industry in Cebu. This industry is segregated into three distinct zones, comprising Junquera, Fuente Osmeña and Uptown Cebu. Each zone has distinctive characteristics: Junquera is the older, diverse and strongly Filipino district; Fuente Osmeña is a newer district catering to white, Western tourists; and Uptown Cebu is the newest district, catering primarily to a Japanese clientele. I present this material with what Gregory (1994) would term 'cartographic anxiety', since I do not wish to present the sex industry as a transparent, mappable reality. Instead this geography emphasizes its heterogeneity, its global links, and its status as a geopolitically contested site which is helping to constitute new social and political identities in the Philippines.

Junquera Street

Although Cebu's residents refer to the red light district as 'Junquera', the actual district spans a much larger area with Junquera Street as a focal point (see Figure 2.1). The vernacular which associates Junquera and commercial sex reflects the area's reputation as a red light district since the earliest years of this century, when the area marked the edge of urbanization and served as a spill-over area for migrants (Mojares 1993). As the city's development proceeded northwest, however, Junquera was largely by-passed by socio-economic development. The area is now home to 4,000 of Cebu's urban poor.

The contemporary Filipino red light district is found in Kamagayan, the municipal *barangay* (suburb) bounded by Osmeña Blvd, P. del Rosario, D. Jakosalem and Sanciangco Streets. The area encompasses the largest number and variety of sex establishments in Cebu. In addition to formal establishments, there are also informal *casas* (brothels) in the residential areas of Kamagayan. Freelance or street prostitution is primarily located on Colon, Junquera and P. del Rosario Streets. Dozens of women do reside and engage in prostitution in Kamagayan, but this is primarily on an informal basis, and where they conduct their business is largely dependent on the degree of police harassment. Indeed, public opinion holds that women

Plate 2.1 A go-go bar/cabaret establishment in the Junquera district. The ground floor is a beer hall, the first floor is a go-go or bikini bar, and the second floor is a go-go bar and cabaret (where men pay women to dance with them). These multi-floor establishments are said to be a result of police surveillance, in that the riskier acts occur on the upper floors.

working in the downtown red light district are residents of Kamagayan, a community frequently referred to as a 'haven for drugs and prostitution' or 'sin district'.

The most distinguishing feature of Junquera is its diversity, but there are other characteristics that make the area distinctively Filipino. Restaurants and food stalls serve Filipino foods and nightclubs play Tagalog and Cebuano songs. Furthermore, the sexual services offered appeal more to Filipino men than those in other districts. The diversity of services – such as brothels, massage parlours, bars and street prostitution – cater to men from a variety of social and economic backgrounds, and even within one type of establishment, such as go-go bars, there are different classes of bars ranging from budget to up-market (Plate 2.1).

Fuente Osmeña

During the late 1970s and early 1980s, nightclubs catering to local and foreign men began to open in the area extending from Fuente Osmeña to the Capitol Site. These establishments came under close public scrutiny

Plate 2.2 A strip of bars in the Fuente Osmeña district. These bars, located near the Fuente Osmeña circle, are American, Norwegian and Australian owned.

due to their live sex shows, however, and a number of them were padlocked permanently. During the 1980s, commercial development along Osmeña Blvd saw the gradual dislocation of this segment of the sex industry, and a focus on Fuente Osmeña (Plate 2.2).

The emergence of bikini bars catering to white, Western men coincided with Cebu's emergence as a cosmopolitan city with Western service outlets, as well as an increased number of American visitors due to the war in Vietnam (Cebu's airport was regularly used as an alternative to the American airbase in the north). Their presence continues, however, primarily due to steady flows of foreign tourists and a seasonal but constant expatriate population. The sexual services available appeal to Western tastes and, as a result, the sex industry in this district revolves primarily around the go-go or bikini bar scene. The bar is a place to socialize and watch a dance show, and the women employed here – although they do engage in prostitution – also serve as cultural interpreters, tour guides and, in some cases, prospective wives.

Bars like the ones in Fuente Osmeña figure prominently in stereotypical representations of sex tourism in the Philippines. Images of Western men taking advantage of poor Filipino women in such establishments – whether in Cebu or near the American military bases north of Manila – have been particularly useful in formulating critiques of American colonialism, cultural imperialism and the commodification of Filipino sexuality. While it is important to consider the colonial/imperial history that may have enabled this type of development, it is also true that this focus has helped to increase the profile of Fuente Osmeña's bars. Such a profile, combined with the visibility of white, Western tourists, has led to a focus of AIDS education efforts in the area.

Uptown Cebu

Just as the construction of Osmeña Blvd pulled Cebu's development northwest during the 1970s, the construction of the Cebu Business Park and its attendant infrastructural projects have pulled development northward since the late 1980s. The major streets are now wide roads and are frontage to new mini-shopping malls, department stores, restaurants, hotels, travel agencies and so on. With this development has come the era of the *karaoke* bar (Plate 2.3).

Karaoke bars catering to Japanese men emerged in the late 1980s and early 1990s throughout the uptown area. Their geographical location is somewhat arbitrary, largely because many Japanese tourists stay on the nearby island of Mactan in beach resorts and are brought by tour guides to the bars. *Karaokes* function in much the same way as hostess clubs in Japan (see Allison 1994), where Japanese men select particular women to serve them drinks, light their cigarettes and orchestrate their singing for the night. While *karaoke* owners claim not to be involved in prostitution – the establishment in Plate 2.3, for example, is advertising for 'guest relations officers' – their employees must submit for a weekly check-up at the City Health Department for sexually transmitted diseases, and a biannual test for HIV. This is largely because women employed in *karaokes* frequently arrange clandestine encounters with their customers after hours to supplement their income.

This differentiation of sex establishments into specific 'zones' during the past two decades reveals their segregation along racial, ethnic and socioeconomic lines, and their location within the city has been related to political, economic and social changes in both the Philippines and internationally. That the sex industry should have a spatial component influenced by these factors is not novel, and has been examined in other areas such as the Mexican–US border (Curtis and Arreola 1991) and the UK (Hubbard 1997). I am recounting this archeology of Cebu's sex industry landscape to emphasize that the globalization of Cebu's sex industry is not a new phenomenon of the 1980s or even the 1990s, and how this complex spatiality helped constitute the politics of AIDS in Cebu. At the same time, however, this differentiation also helped to designate women in the Fuente Osmeña district as both victims and agents of the spread of HIV/AIDS.

Plate 2.3 A *karaoke* bar in Uptown Cebu. This establishment was set to open in the Uptown area in 1994. *Karaoke* bars are establishments that contain several large conversation-friendly sofas, and most men enter these establishments in groups.

The City Health Department and the subjection of 'sex workers'

Increased concern over HIV/AIDS emerged when the first case of HIV was diagnosed in Cebu in 1991. The sex industry came under close scrutiny and in 1992 there were many reports in local and national papers about the increasing numbers of 'sex workers' migrating to Cebu due to the eruption of Mount Pinatubo and the departure of the American military bases at Angeles and Olongapo. Due to the concentration of HIV cases detected around these areas, these women were either presumed to be infected, or seen as 'luring' a foreign sex industry and HIV/AIDS to Cebu. Indeed, semiotic links between the dispersal of Pinatubo's ash and HIV infected sex workers were pervasive. The situation worsened in 1993 when the police raided an experimental clinic for HIV and AIDS patients. The clinic was treating seventeen patients, mostly Americans and Australians, which led the mayor to suggest a 24-hour test centre for anyone who had contact with 'foreigners' in the preceding weeks. The notion that HIV/AIDS was directly linked to a foreign presence in Cebu was widespread, primarily because most cases had been detected in women in the sex tourism industry (although it was rarely mentioned that only these women were tested on a regular, mandatory basis).

The City Health Department has managed the control of sexually transmitted diseases since at least the 1930s, but concern with HIV/AIDS gave them a high public and media profile in the 1990s. According to their statistics, in 1994, Cebu's sex industry was composed of 96 establishments employing approximately 1500 women (Cebu City Health Department 1992). These establishments are further divided into eight different types: *karaoke* bars, bikini bars, singalongs, cocktail bars, beer houses, dance halls, massage parlours and brothels. How this typology came into common usage was not easily discernible, but the typology itself is of interest. The role of the City Health Department is to prevent, control and manage sexually transmitted diseases in Cebu. In Foucauldian terms, and through utilizing his ideas about regimes of disciplinary control, the typology of these establishments could be seen as disciplining the bodies and spaces of commercial sex – simultaneously bringing the 'commercial sex industry' into the sphere of govern-mentality (cf. Foucault 1977, 1978).

While the City Health Department's typology is a project of managing sexuality, delimiting commercial sex establishments as places outside appropriate moral/healthy behaviour in Cebu, it is also a different kind of map – one which relies on the distancing and deterritorialization so common in epidemiological discourse. It is not obvious, for example, that only women are required to attend the clinic on a mandatory basis. They are subject to government intervention because they are most clearly 'out of place': women in the sex industry are often single mothers from urban poor neighbourhoods, and transgress the bounds of an imagined Filipino womanhood. Their presence in tourists districts, particularly in the evenings, marks them as 'bad girls'. Furthermore, and this was a more recent theme, international attention focusing on the spread of HIV – including my very presence in the clinic – incited Health Department staff to cast a suspicious eye on women working in establishments catering to foreign men. This might help to explain the distinction between *karaokes*, which are frequented by Japanese tourists, and *singalongs*, which are patronised by local Filipinos, although it is also true that suspicions about unhealthy/dangerous sex in Fuente Osmeña were more intense. Indeed, such mistrust was evidenced in clinic gossip about the return of American military base workers, including specific information about the neighbourhoods in which these new Fuente bar workers lived. This information circulated quickly, inspiring women working in Uptown and Junquera districts to be suspicious of these women as well. It was also evident in formal discussions about blood testing, particularly in terms of the Health Department's right/authority to test sex workers for HIV. Despite their mask of impartiality on the issue – that is, that *all* workers would be tested – sex workers enlisted to participate in additional HIV/AIDS education activities were often from the bars in Fuente Osmeña. Indeed, HIV/AIDS became an 'epidemic of signification' (Treichler 1989), where discursive epidemiology helped to constitute a geography, with the production of different meanings at different sites. Fuente Osmeña was clearly a site of desire and danger, with the potential to affect the general population of Cebu.

It is important to highlight that this cartography attaches people to stigmatized spaces and bodies – bodies which are frequently marked by sexuality, class and race – tracking the contours of where and who might be 'dirty', 'diseased', 'polluting' or 'sick'. Furthermore, while these conceptions do bear the traces of global AIDS prevention strategies – particularly the transnational usage of the term 'sex worker' – they simultaneously reveal how the concepts of global-foreign-AIDS were conflated, and presumed to be infecting a local, healthy population. In other words, an epidemiological focus helped to ascribe a dangerous sexuality to women who engage in sexual relations in the Fuente Osmeña district. It is in this sense that Cebu's AIDS politics provide an inversion of Stoler's (1995) account of the constitution of bourgeois, Western sexuality. Stoler (1995) contends that Western sexuality and desire have historically been constituted in relation to the colonial other/body, suggesting that the discursive fields in which sexuality emerges can be traced to imperial landscapes. A similar case could be made for City Health perspectives on the sex industry in Cebu, although sexuality and desire articulate with a different colonial other/body – the Western sex tourist – and the landscape is a contested zone of development.[3]

Imperial landscapes: NGOs map the sex industry

The City Health Department may have bureaucratized the existence of commercial sex, but non-government organizations (NGOs) in Cebu have a very different conception of what constitutes the sex industry. NGOs in the Philippines deal with socioeconomic and political issues revolving around development priorities, and have been a cohesive political force since the institution of martial law by President Marcos in 1972. Funding for NGO development projects has historically been sourced from Western donors, and these projects might 'target' the urban or rural poor, women, minority groups, and so on. Instead of writing a typology over the establishments on a map, however, many of Cebu's NGOs are more concerned with nationalist debates that depict commercial sex establishments as inscriptions of foreign control in the landscape.

Cebu is the second largest city in the Philippines and has a metropolitan population of approximately one million people. Since the mid-1980s, Cebu has experienced rapid economic growth and NGOs have been keen to point out that this growth has not necessarily been equally distributed, nor led to the improved livelihood of the local population. Indeed, NGOs frequently juxtapose this phase of rapid and uneven development with President Ramos' plans to integrate further into the global capitalist economy and make the Philippines into a Newly Industrialising Country (NIC) by the year 2000 (Law 1997). Cebu became something of a pilot project for this agenda, and now has an export-processing zone with forty-three jointly or fully foreign-owned firms (producing electronics, handicrafts and processed raw materials), and an international airport receiving direct flights from Australia, Taiwan, Hong Kong and Japan.

Tourist development is often cited as being particularly detrimental to the local population, and many organizations blame tourism, combined with the uneven distribution of economic development, for the size of the sex industry in Cebu (de la Cerna 1992; Dioneda 1993). This perspective meshes well with the nationalist-feminist movement in the Philippines which emphasizes the role of American colonialism/imperialism and debts to the International Monetary Fund and World Bank in perpetuating the existence of sex tourism (Enloe 1989). Yet within this depiction the landscape of prostitution is homogeneous, and commercial sex establishments – whether in Fuente Osmeña or Junquera – are locations on the map of neo-imperial capitalism. The map of the sex industry for NGOs is therefore a signifier of foreign control. Indeed, the spread of HIV was easily assimilated into the politics of anti-imperialism, and AIDS was perceived as another product of a foreign-controlled economy, but one with the potential to affect (i.e. infect) all Filipinos.

Just as the City Health Department maps people into their bodies and spaces, however, so too does the mapping of NGOs. NGO critiques of neocolonial power have been instrumental in dislocating official statements about economic growth, but in the process NGOs conflated the inflow of foreign capital with the invasion of women's bodies. In their desire to highlight the colonialist tendencies of global capitalism, and to advocate changes in local policies that encourage global/foreign influences, NGOs universalized Fuente Osmeña's bars as typical of commercial sex transactions in Cebu. Despite this homogeneous landscape, and notwithstanding their relatively small numbers, two bars in Fuente Osmeña were chosen for NGO seminars on the prevention of HIV (which included briefings on the politics of prostitution and AIDS in the Philippines). Such tactical intervention in the bar scene further reveals how women working in the district of Fuente Osmeña were considered particularly vulnerable/dangerous, helping bound the district as one of virulent sex. Bar owners and employees were mindful of their symbolic role in facilitating such global/local 'intercourse', and there were fears that Cebu's mayor would take precedent from Manila and enforce the closure of Cebu's bars. While this is a fairly simplistic analysis which backgrounds the development of an NGO perspective – it was an anxiously repeated portrayal since its instability was being challenged – it also helps to demonstrate how NGOs utilized a global/local framework to constitute a popular geography of commercial sex.

Located difference: the critique from Manila

Discussions in Cebu did not occur in isolation, however, and eventually came into conflict with best practice models of AIDS prevention which saw them as constructing victims, increasing blame and stigma, and being ineffectual in slowing the spread of HIV. This critique came from Manila, and through the voices of the WHO, Australian organizations and their Filipino NGO counterparts, and was largely a response to a 'clean-up' campaign orchestrated by the local mayor. Mayor Lim had closed down Manila's famous sex tourism district, and this had forced the sex industry underground or to municipal districts a few kilometres south. While a few establishments remained open on a furtive basis, the police frequently raided them. There was little doubt that women working in establishments catering to Western men were being blamed for the spread of HIV, and people were publicly urged to 'stay away from prostitutes' (Plate 2.4).

This fluid geography of Manila's sex industry was influencing popular and institutional perspectives on commercial sex, as well as the types of AIDS education projects required. NGO seminars in nightclubs were no longer effective in reaching underground workers, for example, and bar owners were reluctant to receive groups with official capacities. Understanding how social attitudes and the criminalization of prostitution intertwined with stigma and blame, and how this played a role in the closure of Manila's red light district, inspired new perspectives on both prostitution and AIDS. A National Committee composed of NGOs, the WHO and the Department of Health was formed to deliberate future AIDS prevention strategies, and interrogated the links between mandatory blood testing, visibility and blame. NGOs also reconsidered traditional representations of women in the sex industry as victims of colonialism, imperialism and AIDS, and how this might further marginalize these women. Because a similar 'clean-up' campaign was foreshadowed in Cebu, Manila officials and activists encouraged the Cebu Health Department to rescind its mandatory blood testing policies, and suggested that NGOs redesign their one-way education activities to less judgemental, community-based education projects such as 'peer education'.

Plate 2.4 Manila's sex tourism district 'cleans up'. A sign posted in Ermita, Manila's famous red light district.

Cebu health officials saw the Manila agenda as being Western-based and infringing on their authority, and NGOs saw the recommended 'peer education' models of AIDS prevention as peripheralizing important feminist and nationalist issues in Cebu (e.g. the gendered effects of international tourist development). The peer education model – which had been derived from Australia's gay community and supported through Australian development aid – was displacing the traditional sorts of sex industry projects which emphasized consciousness-raising seminars and job skills retraining to encourage women to leave the sex industry.[4] Contemporary global imperatives were shifting the focus of NGO activities throughout the country to AIDS education, however, and analyses that emphasized the political economy of prostitution and/or the determined nature of the lives of sex workers were becoming increasingly unpopular. Because this perspective influenced funding from international agencies, organizations that used participatory models of education were more likely to receive funding.

Encompassed within these new models of education were recent developments in Western health intervention projects: that is, the project must be designed, implemented and evaluated by the target community, and through this participatory process, the community is empowered to safeguard their own health. While the use of the term 'community' is problematic for sex workers in the Philippines, since it presumes a shared sexual identity, it is also true that these representations were dislocating portraits of prostitutes as victims. This model was also successful in its word-of-mouth, co-worker style, since peer educators could reach workers who were underground or in different bars in the city. While Cebu's activists might be correct in claiming that this figuration is unhelpful in addressing development issues in Cebu, and some might label 'peer education' an example of a colonizing discourse, it is also true that these new approaches encouraged women in the sex industry to take a more active role in the policies and health issues affecting them, thereby subverting the leftist politics which left them with no choice – or voice.

It is in this sense that the Cebu/Manila division was symptomatic of broader changes occurring in the Filipino Left. Magno (1993: 1) has suggested that political movements in the Philippines have become divided into those who 'continue to abide by the hierarchical, statist, and coercive vision of the political alternative' and those who 'view democratisation of the social order as well as political means as the central motif of progressive popular movements'. In other words, there is division between those who feel political processes have not been liberated from their colonial foundations (nor have they improved since the overthrow of the US-backed Marcos dictatorship in 1986), and those who feel that reintegrating the efforts of NGOs with a post-Marcos government is the key to more equitable development. In somewhat crude terms considering the nature of this 'dangling conversation', there is tension between emphasizing how foreign/global influences govern Filipino society – whether this be in the form of foreign investment or support by the World Bank or WHO – and the need for democratic responses to issues which are simultaneously local.

In many ways these positions parallel Cebu/Manila differences. Bureaucrats and NGOs in Cebu conceived Manila's AIDS prevention strategy as a Western solution to a Western problem, and in Manila these problems and solutions were about the creation of a civil society which allowed women in the sex industry to have a role in the prevention of AIDS. While both perspectives are partial since they recount two different sites of commercial sex – with different views and voices in the debate – they also highlight the need to envision a Filipino society that is simultaneously local and global, as well as enmeshed in global structures which have colonial/imperial genealogies. Indeed, Cebu and Manila resonate with a 'global sense of place' (Massey 1994) where sex industries are a locus of social relations complicated by the precarious existence of sex tourists in Cebu and the role of the WHO in helping to formulate national policy in Manila.

Understanding peer education as a colonizing discourse would entail viewing it as a set of values, meanings and practices through which a Western Self and its Filipino Other are defined through an ambivalent cultural politics of domination. In this sense, the wholesale adoption of Western perspectives on appropriate AIDS education activities which do not address concerns about economic marginality could be seen as disavowing global processes of inequality as well as the participation of Western men in the sex tourism. At the same time, however, the production of Cebu's sex industry map was encouraging stigma and blame, particularly for women working in the Fuente Osmeña district. While a Manila-style 'clean-up' campaign did not materialize in Cebu, there were inclinations towards scapegoating Cebu's sex workers for the spread of HIV/AIDS. Furthermore, Cebu's map also reveals a somewhat unanticipated finding: the existence of a large, Filipino-based sex industry. This is unexpected since political mobilization is usually silent on the issue, and Cebu's cultural politics of AIDS was based on representing sex workers as a group of women who engage in sexual relations with foreign men. Yet the production of Cebu's map, together with debates about mandatory testing and models such as peer education, reveals that the sex industry is simultaneously global *and* local, and about economics *and* culture. It is precisely the social and political processes engaged in by bureaucrats and NGOs, where struggles are against – but are also somehow beyond – the ideology and practices of imperialism, that help to frame different, postcolonial questions about sex work.

Mapping Cebu's sex industry

> there is a sense that the Philippines is as much a sovereign nation in the global imaginary as much as it is a series of relations anchored to crisis and contingency.
>
> (Rafael 1995: xvii)

Rather than focus on the determining factors of global AIDS prevention strategies, or the hegemony of WHO-based rhetoric, it is important to emphasize the more ambivalent forms of bureaucratic and activist discourse that mobilize around the place of prostitution and AIDS within a globally-integrated, Filipino society. This approach has much affinity with postcolonial perspectives that seek to disrupt the nationalist pedagogies that set up a binary relation between First World/Third World, and depart from recurrent theories of underdevelopment and dependency. To define peer education as a colonizing discourse, for example, would be at the expense of understanding the Filipino translation and use of the concept (i.e. as a strategy for re-thinking prostitution and coping with Manila's mobile and underground industry). This approach is not apolitical, and emphasizes how a spatially informed perspective on commercial sex enriches our understanding of strategies to prevent the spread of HIV/AIDS. In examining how the geography of the sex industry and bureaucratic/activist discourse were mutually constitutive in both Cebu and Manila, and how AIDS prevention strategies took on different meanings at different sites, I emphasize geographies of power and resistance which combine binaries in unfamiliar ways. These are not merely geographies of WHO power, where the WHO mandates particular AIDS prevention strategies, or geographies of NGO resistance to foreign capital/influence. They also reveal how NGOs have the power to constitute a geography of commercial sex that induces stigma and blame, and how bureaucrats can resist their own, epidemiological geographies.[5]

A map of Cebu's sex industry, while it may signify the global spread of HIV/AIDS, also belongs to a process of narrative description where the politics of place and identity help to define the predicament of the Filipino nation. As a country which is postcolonial in terms of territorial decolonization – particularly since the departure of the American military bases in 1992 – and where national identities of political allegiance are being reshaped by 'global issues' such as HIV/AIDS, the problems of this postcolonial present, and how to address them, are anything but transparent. Jacobs (1996) has argued that the productive encounter between debates on imperialism and postcoloniality can be understood through the specific spaces of the city, and through sites which are saturated with the politics of transformation. The sex industry in Cebu is one such site; this is where the colonial past inhabits but never quite dominates the contemporary politics of place, and where Filipino society can be understood as never being separate from colonial history, while at the same time being something different from it.

Examining the ambiguity of national identity in the Philippines, while at the same time linking these identities with the spatial scales in which cross-class, cross-race, cross-cultural identifications constitute themselves, is one way to reconceptualize debates on prostitution, sex tourism and HIV/AIDS. It becomes possible to examine how the sites of sex tourism (i.e. bars, red light districts, cities, entire countries) are differently defined and understood not only by health officials, activists and international organizations, but by bar women themselves. Furthermore, it considers how the ideological divisions and differences between these people/groups might be revalued, making different relationships possible. This is beyond understanding the spatial technologies of power which discipline commercial sex, and beyond a fixed notion of location which circumscribes oppositional identities such as colonizer/colonized. Pile (1997: 28) has argued that:

> location has more to do with the active constitution of the grounds on which political struggles are to be fought and the identities through which people come to adopt political stances, than with the latitude and longitude of experiences of circumscription, marginalisation and exclusion.

In the Philippines these grounds and identities are constituted through specific sites in the city, through political identities which are enmeshed in debates about colonial history and civil society, and through recognizing the multiple and intersecting relationships which simultaneously place people in positions of power and disadvantage.

This approach to understanding the cartographies of desire which constitute Cebu's sex industry unsettles the notion of a pre-given place called a red light district, demonstrating how purely local or global perspectives are partial at best, and harmful at worst. At the same time, however, it is important to acknowledge the specific arrangements of power which allow local and global perspectives on the sex industry to coexist. Sex industries are contested spaces and, as evidenced by the relations between Manila and Cebu activists, their production increasingly refigures binaries such as local/global or West/East. It is through the emergence of debates which are beyond binaries that it is possible to re-imagine the sex worker as different kind of subject – that is, outside a framework which dichotomizes desire, placing her in symbolic opposition to capitalism and imperialism – and to hear the perspectives of those who best know how to protect themselves from HIV: women working in the sex industry in Cebu.

3

NEGOTIATING THE BAR

Sex, money and the uneasy politics of third space

Stereotypical representations of sex tourism in Southeast Asia have abounded for the past three decades. Images of middle-aged Western men debauching adolescent Asian girls in the red light districts of Patpong in Bangkok or Ermita in Manila, or of servicemen 'letting loose' in the R&R districts surrounding American military bases have inspired extensive critiques of American colonialism, cultural imperialism and the commodification of Asian sexuality. It is difficult to imagine these commoditized sexual relations without at least contemplating the structural inequalities and patterns of globalization which have enabled the development of sex industries catering to foreign men. Indeed, it has been through the important interpretive frames provided by feminist, nationalist and anti-colonial accounts of sex tourism that the stereotypical encounters between 'voyeuristic' Western men and 'submissive' Asian women have gained a politically strategic coherence.

The popular representation of go-go bars as sites of foreign oppression underpins, and plays a major role in sustaining and authenticating, political mobilization against sex tourism in Southeast Asia, as well as internationally. Within the language of advocacy, prostitute identity is metaphorically fixed within a rich-Western-male/poor-Asian-female dichotomy, ultimately conveying a powerful subject/disempowered other. While this representation plays an important role in highlighting the economic, political and social bases of inequality, it simultaneously reinforces the hegemonically constructed identities of the 'oppressor' and the 'victim' through naturalizing them as fixed identities and subject positions. In so doing, it offers little room to manoeuvre, to negotiate identity or to resist the complex power relations constructed at points where class, race and gender intersect.

The aim of this chapter is to destabilize the naturalness of the encounter between Western men and Southeast Asian women through an analysis of space, and more specifically, to deconstruct and reconstitute the places of sex tourism as negotiated spaces of identity. This is not to deny the relations of power in the sale of sex; it is a rather different conception of power which enables us to conceive more nuanced geographies of resistance. In this way, it can be suggested that the white, male gaze is not merely an autonomous voyeurism – it has its own difficulties and uneasiness due to the gaze from the supposedly powerless bar women, who have their own sights/sites of power, meaning and identity. It can also be conveyed that bar women are capable of positioning themselves in multiple and intersecting relations of power – which include, but are not exclusive to their encounters with men – and that it is in these spaces that subjectivities, capable of resistance, are forged. By analysing the space of the bar in this manner, the presumed identity of women *as prostitutes* and men *as customers* is called into question, as well as the moralizing discourses upon which these social/political portraits of oppression depend.

To examine the places of sex tourism as negotiated spaces of identity, I situate this conception within a particular space: a bar in Cebu City.[1] The bar is owned by an Australian, managed by a Filipina, employs approximately 20 Filipina women, and is primarily frequented by white, Western men. Despite its vernacular status as a 'foreigner bar', however, its space is far from being clearly indigenous or foreign; it is neither and both depending on how it is framed and experienced. Nor, as many pro- or anti-prostitution activists would argue, is the space of the bar clearly liberating or oppressive; it simultaneously offers the possibility of both liberation and oppression, together with a range of other experiences. While there is no 'true' reading of this landscape, the space itself is not entirely innocent. For the space of the bar is also situated within other spaces of political, religious and moral significance, and these conflicting and interpellative discourses play a role in how this real-and-imagined space is experienced and understood. Indeed, the bar is the place where the real and the imagined merge to mediate the performance of identity.

Because Cebu's bars, and the encounters occurring within them, are beyond dualistic economies of meaning and power, I begin by specifying an understanding of 'third space', as well as its implications for strategies of resistance. This is followed by a narrative of the everyday social relations in a bar in Cebu City, mapping a space for a Filipina subject. The bar is a space where dominant images and stereotypes are contested, where people speak from spaces beyond conventional representations of sex tourism, and where resistance, in potentially subversive forms, is possible. At the same time, however, it is the uneasy politics of third space encounters which invite the performance/visibility of fixed identities – identities which are thoroughly intelligible though representations which simultaneously circumscribe the stereotypes of the 'oppressor' and the 'victim'. I conclude by asking questions about the constitution of subjectivity in space and the spatial constitution of subjectivity, and how an understanding of third space can aid in our conception of geographies of resistance.

The ontology of this approach is not apolitical; instead, elaborating the negotiation of power and identity – revealing ambiguity, displacement and disjunction – refuses to totalize experience and therefore offers the potential to locate where more subtle sites of resistance are enunciated. There have been many ways of articulating the 'problem' of sex tourism within various political discourses in the Philippines, but these discourses remain largely dominated by the interpretations of middle-class activists. This is at least partially due to the reality that comprehending the political, economic and colonial dimensions of sex tourism does not necessarily provide a basis for understanding the more personalized modes of identification which form around issues of race, gender and sexuality. In short, emphasizing the determined character of prostitution ironically tends to alienate its subject. Elaborating a negotiated space in order to find contemporary sites of collaboration and contestation and new perspectives on identity, power and resistance provides a means to imagine more ambivalent deployments of power, and how space is constitutive of this process. For what if meaning and power are not necessarily mobilized through stereotypes and their associated dichotomies? Culture generally, and strategies of power in particular, are not found in predetermined categories of identity and experience, and power has the potential to be mobilized through ambivalence.[2]

'Third space', identity and resistance

> Meaning is constructed across the *bar* of difference and separation between the signifier and the signified.[3]
>
> (Bhabha 1990: 210, emphasis added)

Drawing on recent writing in feminist and postcolonial studies, Pile (1994: 255) has argued that dualisms are 'intended to mark and help police supposedly fixed, natural divisions between the powerful and the disempowered', and that 'if we accept these dualisms then we collude in the reproduction of the power-ridden values they help to sustain'. As dualisms are more fluid than their architecture would suggest, Pile outlines an alternative to dualistic epistemologies which incorporates the notion of a new geometry of knowledge, or a 'third space'. This space is a location for knowledge which: (a) elaborates the 'grounds of dissimilarity' on which dualisms are based; (b) acknowledges that there are spaces beyond dualisms; and (c) accepts that this third space itself is 'continually fragmented, fractured, incomplete, uncertain, and the site of struggles for meaning and representation' (Pile 1994: 273). An epistemology which uses the concept of third space therefore encourages a politics of location which recognizes the 'social construction of dualisms as part of the problem', as well as 'places beyond the grounds of dissimilarity – collectively named the third space' (ibid.: 264).[4]

This conception of third space is useful for examining the articulation, transgression and subversion of dualistic categories; indeed, in a metaphorical and material sense, this third space is the bar. The bar is the location of difference, particularly of the cultural, racial and sexual differences between white, Western men and Filipino women. It is, therefore, the location where various bar experiences are articulated and assigned meaning by bar employees, the men who frequent these establishments, and researchers such as myself. Yet this third space does not contain preconstituted identities which determine experience, nor does it possess an authentic character or identity. Instead, identities are continuously negotiated through this space of difference – apropos Bhabha's (1990) opening quote to this section – and therefore constituted through encounters with otherness. To elaborate this concept of negotiation, Pile draws on Bhabha's notions of hybridity and identification. For Bhabha:

> the importance of hybridity is not to be able to trace two original moments from which the third emerges, rather hybridity . . . is the 'third space' which enables other positions to emerge. This third space displaces the histories that constitute it, and sets up new structures of authority, new political initiatives, which are inadequately understood through received wisdom.
>
> (1990: 211)

Moreover, in enabling such other positions, this third space is therefore:

> not so much [about] identity as identification (in the psychoanalytic sense) . . . identification is a process of identifying with and through another object, an object of otherness, at which point the agency of the identification – the subject – is itself always ambivalent, because of the intervention of that otherness. But the importance of hybridity is that it bears the traces of those feelings and practices which inform it, just like a translation, so that hybridity puts together the traces of certain other meanings or discourses. It does not give them the authority of being prior in the sense of being original: they are prior only in the sense of being anterior. The process of cultural hybridity gives rise to something different, something new and unrecognisable, a new area of negotiation of meaning and representation.
>
> (Bhabha 1990: 211)

I quote Bhabha at length due to the complexity of his statements, drawn – as Pile (1994) observes – from the psychoanalytic and post-structuralist traditions. What is stressed here is the ambivalence, and not fixity, of the construction of identity. 'Prostitutes' and 'customers' are actively produced through inherently unstable social encounters, and it is through these intersections of power and difference that they locate their oppositional identities. Both bar women and customers, for example, are conscious of the moralizing discourses of sex tourism, yet their constructions of prostitution draw from various and different cultural, historical and gendered positions. While some Western men might be inclined to construct the issue around the moral and secular perspectives of Western feminism (e.g. the subordination of women), Filipinas might be more inclined to situate their employment within the struggles of everyday life and Catholic beliefs (e.g. suffering, shame and martyrdom). Yet the result of negotiating these differences may be the articulation or subversion of prevailing notions of powerful/powerless in terms of how they experience their encounters: a 'submissive' Filipina may simultaneously be striving for self-actualization, a 'voyeuristic' Western customer may be humbled by an awareness that his masculinity is yoked to his pocket book. The third space is therefore capable of disrupting the 'received wisdom' which portrays sex tourism as an uncomplicated relation of domination.

Homi Bhabha's 'interstitial perspective' has most recently been criticized by feminist geographers as a disembodied – and gendered – perspective (Rose 1995). While a feminist critique of Bhabha is beyond the scope of this chapter, it is important to acknowledge at least some of these criticisms. First, Bhabha's attempts to disrupt the dominant gaze in colonial encounters spoke to my understanding of the relationships between men and women in Cebu's bars (which were simultaneously about race, class and sexuality). Bhabha's perspective appeared to be a useful way beyond the stereotypical representations of the voyeur/victim, opening a space for alternative subjectivities capable of differentially engaging the relations of the bar. Second, and if our theories are to be embodied, then hybridity and third space reflect my position as a researcher, as a white, Western woman who simultaneously identified with both *Western* men and Filipina *women*. It was not possible to locate myself at either the margins or the centre in this space – such a move would be dubious privilege. Third, while Bhabha's conception of third space stands accused of lacking critical possibilities or a radical potential, what I attempt here is to develop Bhabha's project in ways which take these concerns into account.

If the bar is a third space, an ambivalent space of negotiation, and a site of struggle for meaning and representation, then this approach to understanding power and identity poses interesting questions for geographies of resistance. In the context of Cebu's bars, these questions circulate around issues of how Filipino women find ways to engage Western men in ways which are not outside power, but are in interstitial spaces between power and identity. By dancing on the bar, and negotiating interstitial spaces through their own perspectives, women resist the power of the voyeuristic gaze through disruption rather than covert opposition. Yet this approach to resistance necessitates a different conception of power – one that is transient, flexible and ambivalent. Writing on resistance has traditionally relied on an assumption of power which was thought to oppress coherent human subjects. Resistance was therefore anchored to an essential measure of agency (usually class) as a point from which to undertake political projects.[5] The theoretical displacement of the human(ist) subject has most recently inspired critiques of these studies, which are said to neglect the complex layerings of meaning and subjection which inform social encounters, as well as how the deployment of power is frequently laced with contradiction, irony and compromise (Abu-Lughod 1990; Kondo 1990). Indeed, Kondo (1990: 219) has asked if 'articulating the problematic of power in terms of resistance may in fact be asking the wrong question'.

If everyday sites of struggle occur in places of overlapping and intersecting forms of subjection, and if domination is never achieved without ambivalence, then how are we to envision power and resistance? What is evident is that Foucault's (1978: 95) formulation 'where there is power, there is resistance' is no longer sufficient. One way forward is Abu-Lughod's (1990) reformulation 'where there is resistance, there is power', and resistance is understood as a 'diagnostic' of changing relations of power. Although Abu-Lughod's focus is on gender relations, by analysing resistance on its own terms, she displaces abstract theories of power with subjectivities capable of contingent and flexible modes of resistance. Through focusing on these resistances, we can begin to ask questions about precisely what relations of power are at play in different contexts. This approach therefore addresses the problem of attributing people with consciousness which is not a part of their experience, as well as opening the possibility for examining the more localized contexts of resistance. Within the internal politics of the bar, for example, bar women are not merely resisting white, Western men; they are resisting stereotypical encounters with men, their co-workers and researchers such as myself, where the categories of race, class, gender and sexuality place them in uneasy positions of coherence.

Now I introduce a new border – a shift to a narrative voice – to explore this conceptualization further. Drawing on the critical style of Malcomson (1995), I strategically destabilize the boundaries between self and other by replacing a first person narrative ('I') with a second person narrative ('you'). Malcomson's essay on the multiple identities performed by customers in a bar in Bulgaria formed the inspiration for this narrative, particularly his emphasis on the fluidity of bar patron identity. The second person narrative is not meant to reject or dismiss my own gaze, but rather to decentre and destabilize it – allowing 'you' to imagine and experience presumably discrete identities in different contexts.[6] It is also a device of memory, where the academic 'I' remembers the fieldworker (in this case 'you'), and a modality which reflects my own resistance to the confines of an academic discourse which occludes the smoke, beer, laughs, intimate chatter and secrets which permeate the space of the bar.

The dialogics of the dance: an evening at the Brunswick

It's 5:30 p.m. and you're in the Brunswick waiting for Cora. She agreed to meet for a beer before work so you could talk about her life in Cebu. Last month she told you she was completing high school in Leyte a few months ago when her poverty-stricken parents forced her to migrate to Cebu to seek paid employment. Because of her lack of education, she resorted to working in a bar: 'I had no choice,' she'd said, 'what else could I do?' You know this is a common story in and around the growing metropolis, but Edna – another employee of the Brunswick and your friend for almost a year now – said they worked together in a bar around the American military bases before the Mount Pinatubo eruption. 'That's why she got those breast enlargements', Edna elaborated, 'and where did you think she learned to speak Pampangan? Of course she's been working in a bar before.' You're interested in hearing Cora's side of the story.

It's 6:30 p.m. and you sit at the bar watching women trickle into the Brunswick from the crowded streets of Fuente Osmeña. Anna, Edna, Maryann and some new women have arrived, but there's no sign of Cora. While you wait, you spend some time talking to Fely, the manager, who has proven to be an endless source of anecdotes about life in the Philippines. Most of the bar women believe Fely used to work as a prostitute in Manila, where she met her husband, the Australian owner of the bar. Fely disregards such *tsismis* (gossip) as pure jealousy over her elevated social status, however, and tells you grand stories about working in various jobs around the Philippines.

Fely spends time talking to you because she too is taking note of recent arrivals; indeed, she is waiting for them. Each night she sits at the bar writing each woman's name onto a list as they arrive. The list is very much like a roll call – Fely is very strict on attendance and punctuality – but will also serve as the dancing order and an inventory of the ladies drinks each woman earns for the evening. Fely's air of formality and strictness contrasts with the chatting and laughter emanating from the back rooms where her employees are spending their time preparing for the night: dressing up, putting on make-up, eating dinner, chatting about last night's business or their boyfriends and wondering if the night's business will be good.

Cora shows up at almost 7:00 p.m., and you know she doesn't have time to talk. She just strolls past by you, barely catching your eye and not saying a word. Later Edna will inform you that Cora cannot separate your identity from that of a religious or

> social worker, and after dealing with your frustration – you've emphasized your support and don't want people to hold that opinion of you – you accept that as a good strategy for women who have been previously condemned for their employment. 'Anyway,' Edna will say, 'she'll never understand and you have lots of friends who want to talk to you in the bar already.'

This passage is simultaneously a story about identity, power and resistance. Cora's refusal to separate my identity (as a student/researcher) from a religious or social worker[7] can be read in many ways. First, it can be seen as resistance to confronting the power of moralizing discourse; Cora resisted such power through refusing to talk. Second, Cora's identification of herself as a particular subject – an immoral Filipina – relates to prevailing attitudes about prostitution. Most bar women are aware that in many Western countries, as in the Philippines, prostitution is conceived of as an immoral form of employment, although Western men and women are more inclined to understand the context of poverty in the Philippines. While this identification may have been ambiguous for Cora, it nonetheless impeded our ability to communicate. Edna, on the other hand, was pleased to inform me that I had many friends in the bar already, and that my research did not require a formalized interview with Cora. This identification – where to Edna my interest developed from research to empathy and friendship – led to a new area of negotiation between friends like Edna and myself, where I learned that constructing her as a subject was only possible through my academic pursuits. To Edna I was merely a student at the local university who was interested in the lives of bar women and, while she patiently answered my many questions, I assumed part of that identity too.

Cora's breast enlargements are one example of how the bodies of Filipina bar women – or more appropriately, 'dancers', as they refer to themselves – become marked through the selective incorporation of features of an imagined Western sexuality. Other examples would be the use of cosmetics, and various drugs to control their weight. These practices are often interpreted by middle-class Filipinos as a purely Western influence – large breasts and skinny bodies are not highly valued in the Philippines, but are dominant in imported pornography and television programmes – but they could also be read as the psychological dynamics of resistance which allow women to retain control of their bodies. Dancers re-learn their bodies in new and often revolutionary ways during their employment in the sex industry, and this is often equated with 'becoming modern'. Yet while dancers' bodies express modernity they also bear the marks of being Filipino. Their dancing styles and entertaining costumes, for example, follow distinctively Filipino fashion trends. It is precisely this hybridity – exotic otherness and a sexualized modernity – that male customers are often attracted to (cf. Manderson 1995).

Finally, it is important to note Cora's assertion of having 'no choice' but to work in a bar, which bears resemblance to Foucault's (1978) notion of a 'reverse discourse'. In the Philippines, political mobilization against sex tourism has primarily been based on analyses which situate prostitution within the political economy of colonialism, militarism and sex tourism.[8] By emphasizing the structural determinants of why women enter the sex industry, however, Filipina dancers are cast as 'victims' who have no choice. While it could be argued that this representation victimizes women, denies agency and distorts the complexity of experience, here it is important to note that the naturalness of the choice issue can be appropriated by women themselves, where having 'no choice' simultaneously becomes a source of agency, a resistance to moral judgements and justification for their employment.

> It's Maryann's night to do manicures so you go out to the back room to get one yourself. Maryann is very petite and shy, and is fussy about the customers she goes out with, so Fely has agreed she can earn extra money from this side-business one night a week. This used to be Maryann's regular employment in her *barangay* (suburb) until her husband was sent to jail last year and she needed more money to support her two children. Everyone agrees Maryann's manicures are very good, and she needs the money, so you splurge and get your toes done with red polish too.
>
> While you get your manicure, you realize that your own placement in the bar is related to your relationships with the women – it took a long time until you were invited to this back room, for example, or until you were comfortable sitting out front by the dance floor. You also wonder about the recent government AIDS information campaign about sterilizing manicuring equipment between customers, but know this is not the time nor place to mention it. Whenever you've tried to get to this aspect of your research – to discuss AIDS as an issue for sex workers – this space becomes awkward and silent. You've wondered if AIDS has been delimited as an impossible subject, and if there are too many borders for you, as an outsider, to cross. Of course they use condoms whenever they can, but by virtue of the fact that you are not a dancer, it is difficult for you to participate in discussion of the times when they don't. Dancers are obliging in terms of discussing their sexual lives with you, but condom use, if discussed in a rational or clinical way, is difficult. You end up feeling like a City Health Official and the answers you get are rehearsed.

The recent government AIDS education campaign did not go unnoticed by bar women in Cebu. This was partially due to the increased information drive at their weekly check-ups at the City Health Department, but also because there had been several interested non-government organizations through the bar over the past year asking them to fill out KAPB (knowledge, attitude, practice and belief) questionnaires. The most profound effect, however, came from the release of a Filipino film, *The Dolzura Movie*, which was a more emotionally-charged perspective on a woman who had contracted the virus. While bar women were happy to answer KAPB questionnaires and discuss the movie in terms of their fear, they located these discussions in two distinct discursive realms: KAPB questionnaires were part of an official education campaign they were required to participate in, while the film dealt with personal circumstance, emotions and love.

There are several academic and activist 'theories' regarding why women in the sex industry do not use condoms for every sexual encounter. In the Philippines, it is emphasized that women in the sex industry do not use condoms with their regular partners, and while at work, have less negotiating power due to the oppressive nature of the business. While it is true that bar women tend not to use condoms with their regular boyfriends in order to separate work from pleasure, to equate the non-use of condoms with general disempowerment distorts the issue. The non-use of condoms, in what would appear to be unreasonable circumstances according to middle-class sensibilities, has more to do with Filipino conceptions of *bahala na* – translated by Enriquez (1990: 302) as 'risk-taking in the face of the proverbial cloud of uncertainty and the possibility of failure' – than with disempowerment or a lack of negotiating skills. Many women employed in these establishments are looking for a future husband, and the use of condoms in these instances – that is, with a prospective husband – is seen to impede the achievement of intimacy, and ultimately, an opportunity to exit the industry.

Awkwardness, particularly in the form of silences and bodily discomfort, is a common phenomenon in the bar. Such awkwardness occurs in the space between attempting to comprehend a situation, and articulating a (suitable) response to it. In a sense this uneasiness may embody the third space concept, and is what makes the bar a 'special' space. If third spaces are merely spaces where dualisms are worked out, then all spaces in a sense are third spaces. But in the Brunswick, as in most bars, uneasy silences and bodily discomfort are better conceived as the result of uncertain identities in tension. In the back room of the Brunswick, for example, discussions of HIV/AIDS had previously provoked the performance of official, public health identities, but these identities were inappropriate/d for a space which was usually used for chatting and socializing. Rather than interrupting this space with what could be considered an interrogation, I opted instead to listen to stories about Maryann's children, her previous employment and the high quality of her beautician skills.

Although discussions on HIV/AIDS in the back room fell outside appropriate codes of conduct, the public area of the bar is a space where dominant codes can be transgressed. Dominant codes of morality, for example, prohibit commercial sex between Western men and Filipino women. It is the possibility of precisely this kind of sex, however, that creates a third space of bodily discomfort and a quest for an individual subject position. If the third space is a 'space' of negotiated identity, and more specifically, if it is a space beyond dualisms, then both the bar and the body can be conceived of as sites of ambivalence and negotiation. In this space dancers are awkward on their first nights in the business, and new customers tend to be rather gauche. Researchers are 'out of place' in this environment, so negotiating an appropriate subject position – whether it be in the back room discussing HIV/AIDS or sitting out front watching a dancer perform – is hardly surprising.[9]

> A few customers arrive. You look at them and suspect they're probably tourists, and probably here for the first time. They look nervous and unsure of what to do or where to sit tonight, but by tomorrow you know they'll be acting like regulars. You've stopped being so self-conscious and aware of the white, male tourist gaze for some time now, and briefly wonder why they're staring at you. One of them looks as if he is about to engage you in conversation, but Edna comes over to remind you about the birthday party in her *barangay* the next day. You're grateful you don't have to talk to this man after being berated by a drunken Australian the week prior. He had demanded to know why you were in the Brunswick, and when you told him, he accused you of wasting Australian tax money by carrying out research in bars, and particularly because you're Canadian.
>
> Edna's neighbour Lorna, who works at a nearby bar, comes in to see what's going on in the Brunswick. She's on a steady barfine this week; she's entertaining a young Australian 'boyfriend' who's she's been travelling around Cebu with. Last night he took her to a beach resort where they went jet-skiing and skinny-dipping, and today she's lost her voice so cannot work. She and the Australian are playing pool in a nearby pub, the Richmond, and she asks you if you'd like to join them. The Richmond is a staple hangout for foreigners, particularly for local expats and tourists who have come to Cebu to scuba dive. Lorna knows all of the bartenders and waitresses – many of her customers bring her here to socialize – so you stop to chat and compare notes on Lorna's boyfriend. Fely has unofficially banned all Brunswick employees from entering the Richmond unescorted, however, because she says it makes them look cheap because they have no reason to be there apart from

> 'hunting'. Lorna points out the 'hunting girls' that are working the bar tonight. 'Hunting girls', she tells you, 'aren't dancers, and don't work in a bar. They're just working when they want to, and looking for a man who will give her all their money – no barfine – and maybe she'll tell them she's not working, it's just for fun.'
>
> You play a few games of pool and think the Australian is OK. You want to check him out because Lorna is pretty serious about him. This is his second trip to see her, and he sent her twelve red roses for her birthday last month. She confides that she's planning to have his baby, while you're happy she's happy, you wonder what she needs another baby for; she already has two Amerasian children from when she was working in Olongapo. But you say nothing of the sort because you're happy for her, you know how much Filipinas love children, and how dancers often see children as an opportunity to solidify a relationship.

The community where Lorna and Edna, and several other Fuente Osmeña bar women live has an extensive history of women migrating to work in the bars around the American military bases at Angeles and Olongapo. The first bar woman is rumoured to have migrated in the late 1960s, after which time a series of personal recruitments produced a steady flow of women to both cities. Many families rely on remittances from daughters elsewhere in the Philippines, or from daughters who have married American men and are now living in the United States. The Mount Pinatubo volcanic eruption in 1991 saw the unequivocal departure of American servicemen, however, and this event also displaced thousands of dancers in both cities, many of whom came from communities such as Lorna's. Both Lorna and Edna migrated back to Cebu in 1992, and obtained employment in the bars of Fuente Osmeña. In contrast to the perspectives of those who strive to abolish sex tourism, it is important to note that they do not see themselves as victims to American imperialism.

The phenomenon of 'hunting' should not pass unremarked, although my contact with these women was fleeting and occasional. The women I spoke to who were 'hunters' either currently, or had previously, worked in a bar in Cebu. They were frustrated with the bar owners' commissions on their 'barfines', which were perceived to be an unfair reduction to their salaries (primarily thought to be obtained from dancing and entertaining customers in the bar).[10] In this case hunting provides the opportunity to transgress the boundaries of a dancer identity which, as evidenced in Fely's strictness in attendance and punctuality, as well her ban on employees in the Richmond, is also one of 'employee'. Hunting emphasizes women's looking and agency, revealing their ambivalence in negotiating their identities with Western men in different spaces.

It's 10:00 p.m. and you walk back to the Brunswick. Some of the regulars are around and the space feels more comfortable: there are more people around, there is less overt staring and the women are happy because they're earning drinks. One of the regulars is May's German customer Hans, but that doesn't stop a group of women from coaxing him to buy them a drink: Alice is telling him about her sick mother, and Anna is telling him she needs a break from working but her family cannot afford it. Ladies drinks cost the customer from P50 to P100, and the women receive a P25 to P50 commission for each drink. The women know Hans is sometimes generous and, anyway, May has a Filipino boyfriend at home. You look at May, she's acting so confident – she knows she'll be out making money on short time tonight – and you remember that she's almost 35 years old although she looks much younger. She's been working in the business off and on since she was 18, and has lived with two former husbands in Europe and the Middle East respectively. She always gets the German customers because she speaks their language.

You go and join the group and say hello to everyone. The women are chattering in Cebuano and you hear them discussing how cheap Hans is because he won't buy anyone but May a drink. Behind their backs such customers are 'cheap charlies' or '*kuripot*' (in Tagalog, the national language, this is the word for 'cheap'), but Hans has been in the Philippines long enough to speak Tagalog, so they're calling him '*tahik*', which is the Cebuano translation. You ask him if he's buying rounds, but he tells you – for what is probably the fifth time – that he just runs a scuba-diving shop and earns Filipino wages. You laugh but understand his point and so does Edna, who tells everyone to go look somewhere else for a drink. Hans buys the three of you a beer.

You ask May about the women soliciting drinks and she reminisces about a time when things were different. 'These girls we have right now, they're only interested in money,' she tells you, 'like if they can get a customer that night. You see how they react if a customer walks in, and I hate that. Our old group was not like that before. I mean, we wait, give him a chance. Not like now, one customer and five girls at the table, its supposed to be the customer wants her or likes her, you know, and then they cannot say anything because they are too embarrassed to say.' You agree but wonder if a round of drinks isn't really that expensive after all.

While Edna and Hans talk about scuba-diving, you and May notice one of the new dancers on the stage. She looks extremely uncomfortable because tonight is her first night, and May informs

> you she's a virgin. She was recruited by a friend who also took a job in the Brunswick as a virgin but married her first customer, a Norwegian man, after only three days on the job. 'I talked to her tonight,' May tells you, 'and I said if you are dancing, dance with the music. Don't dance and look around because you don't know how you're reacting on the stage. You know, it's boring. The customer, instead of saying "hey she's nice", says "hey she's boring". So I told her that if you dance, don't think about the girls around you. Listen to the music and dance with it, then you will feel good. That's what I did before.'

Calling Hans cheap to his face, but in another language, can be understood as a source of resistance; but it is a resistance directed not only at Hans, it is also directed at May. Hans is an expatriate with a Filipina wife, and only engages in 'short time' with May. Both May and Hans have Filipino partners they return to each night, and neither are interested in an emotional relationship. Because Hans is not a prospective husband for women in the bar, but also because they know his wife – she used to work at the Brunswick a few years ago – many dancers question both his and May's morality. It is frequently insinuated that Western men differ from Filipino men in that Filipino infidelity is more concealed, and therefore more respectable. Therefore, while Hans' ineligibility and infidelity should deter women from soliciting drinks so staunchly, their inclination to solicit May's customer in particular is derived from wanting Hans to pay for his pleasure, on the one hand, while questioning May's morality on the other. Many women justify their employment in the sex industry through their desire for a better life, symbolized by their desire for relationships with customers. It is precisely this absence of desire in May that vexes them.

Although my purpose here is not to analyse 'romance', the centrality of love and emotional pursuits should not be denied. Indeed, May's attraction and 'short time' with Hans are criticized by many women because they are not interested in 'love'. It would be unfair to cast dancers as merely seeking the allure of wealth or an escape from poverty, because their relations with foreign men – even if they desert or disappoint them – are most often seen fairly favourably. Furthermore, Catholic ideas about victimhood and martyrdom play a constitutive role in terms of their emotional desire. It is difficult to place their desire outside the pursuit of happiness via foreign men, marriage and more children, and this desire also connects to the non-use of condoms to achieve intimacy with customers. *Contra* Hochschild's (1983) position on the professionalization of emotions in the hospitality industry, however, there are 'real' relationships that form in the bar, happy marriages are consummated, and these relationships often provide both the dancers and the customers with opportunities for marriage. These contradictions, which are marginalized within current debates on sex tourism, are important in the constitution of dancer subjectivity.

Because of her age and long-term employment in the bar, May usually takes it upon herself to initiate new women through advice on interacting with customers, other bar women, as well as miscellaneous tips on personal hygiene (usually on condoms, personal and customer genital health and how to avoid being diagnosed positive for STDs at the City Health Department). May's advice on dancing in this passage is interesting because she identifies an important gaze as coming from the women themselves (i.e. not the customers). Most women identify their first few weeks of employment in the sex industry as tremendously difficult, and this is at least partially due to the look May specifies. Because the bar is a space that breaches dominant codes of morality, new dancers must negotiate the way this space is understood. As the opinion on May's involvement with Hans indicates, it is also a judgemental look that continues throughout the course of bar work.

> It's almost midnight and a man you might call a holiday regular comes in. His name is Derek, he's a successful businessman in Australia, and he spends all of his holidays in Cebu. You've had the opportunity to talk to him before because his good friend is your current housemate's boyfriend; you've all been for dinner, dancing and hanging around to enjoy the relative luxuries of the Graduate Hotel. The bar is crowded and one of the few seats left is beside you. He comes over to join you, May, Hans, Edna, and what would appear to be Edna's impending barfine. Derek asks you how your research is going and you tell him you're doing fine. He always wants to talk about your research because he's certain that he's an expert whose opinion you should solicit. You do want to talk to him, since his presence plays an important role in dancers' lives, and you recognize that customers all have their own constructions of bar work. Tonight Derek wants to talk about how the bars in Cebu are no different from the pick-up bars in Australia, except that the women in Cebu get paid for their sexual liaisons. He dissociates prostitution in Cebu from prostitution in Australia because in Cebu the women are interested in more than sex, and are willing to spend more of their time with him. 'It's really a much better deal for your money,' he says, 'I've been coming here almost every month for two years.'

Derek's perceived parallel between Cebu's bars and pick-up bars in Australia is partly a denial that he is paying for sex, and therefore, by implication, that he is morally degenerate and sexually inadequate. At the same time, however, it must also be acknowledged that dancers in

Cebu consider themselves more than prostitutes, they are also city guides, interpreters of local culture, prospective wives, and so on. Both dancers and customers maintain these impossible spaces for themselves – as pick-up artists, as prospective wives – spaces that subvert dominant codes of behaviour and morality in interesting ways. Resistance in these spaces is about negotiating uneasy and contradictory feelings and desires.

Furthermore, it is important to note Derek's dissociation of these bars, and his desire for relationships which are about, but are also more than, sex. As mentioned above, customers are often attracted to the hybridity of dancer sexuality, but in some emotional relationships the contradictions of these desires may not be resolvable. While some men desire the sexy selflessness of Asian women, for example, it is also true that employment in the sex industry is constitutive of dancer's self-actualization. In this sense the women's intentions contradict the assumptions of the 'submissive' stereotype, emphasizing that this identity is not accepted without struggle. Yet the desire for a sincere relationship does exist on the part of some customers, a large proportion of whom are divorced and in their forties and fifties. While such desire meshes well with dancers' desires for marriage, it is also true that many of the women do not find many of the men physically attractive. These men are therefore in a position where their allure is their wealth. While this position is a form of power, it also underlines the contradictory relationships between sex, money and the uneasy politics of third space.

Dancing on the bar: mapping geographies of resistance

> [Nietzsche] insisted on a new type of philosophy or knowledge, one which, instead of remaining sedentary, ponderous, stolid, was allied with the arts of movement: theatre, dance, and music. Philosophy itself was to be written walking – or preferably, *dancing*.
>
> (Grosz 1994: 127, emphasis added)

Dancing on the bar, as a play on words, is one way to imagine the subjectivity of Filipino dancers. First, it bears in mind the importance of dancing to employment in Cebu's sex industry. Working in a bar is about, but also exceeds, the exchange of money for commercial sex. Second, dancing *on* the bar is an attempt to displace the dominant representations of sex tourism which, through emphasizing the oppressive relationship between Western men and Filipino women, fails to recognize the production of identities through the negotiation of the sex tourism encounter. Finally, it is a way to stress the importance of the kind of *dancing* to which Grosz refers; that is, a way to recognize that women in Cebu's sex tourism industry are capable of manoeuvring to position themselves within multiple and intersecting relations of power.

The relationship between resistance and subjectivity, however, is more problematic. It might be helpful first to differentiate between identity and subjectivity, demonstrating how the performance of particular identities can itself be understood as resistance, while at the same time recognizing that the performance itself tells only part of the story. In their analysis of Rio Carnival, for example, Lewis and Pile (1996) have suggested that the performance and masquerade of identity (in this case, femininity) can be a form of resistance that renders that particular identity indeterminate and unknowable. Following Butler (1990, 1993), they maintain that the most productive effect of power is to secure these identities as visible/intelligible. Within this framework, it could be argued that dancers in Cebu are aware of, and perform, their status as objects (of Western desire) and subjects (immoral Filipinas). Many incorporate features of an imagined Western identity (e.g. the use of cosmetics, tobacco, and the attainment of an ideal breast size or weight), for example, particularly as they 'become modern'. Furthermore, and within more traditional understandings of performance, women often perceive their sexual relations in emotional terms. Although these fantasies and incorporations only become visible through stereotypes which deny the lived experience of sex work, reading them in this way also has the potential to undermine resistance as either reproducing stereotypes or as participating in their own oppression. It is therefore important to acknowledge that as these women 'become modern' and experience emotional attachments, new images and desires are simultaneously being produced.

These new images and desires exist in a third space of possibilities which are not capable of being understood within current theorizing on sex tourism in the Philippines. At the same time, however, it is precisely these possibilities which are constantly engaging meaning and power: the bar is one site; the body, however, is another. Yet by restricting an analysis of resistance to the performance of identity, demonstrating how dancers ritually escape dualistic economies of meaning and power, dancers are reduced to a series of 'subject effects' which have little internal coherence (Ortner 1995). Instead of focusing on these 'effects', it is useful to focus on what becomes defined as contested terrain, and how resistance can, and does, surface.

Cora's refusal to speak to me, reactions to Alma's engagement in 'short time' with Hans, and Derek's dissociation of prostitution in Cebu are good examples of how power can be produced through stereotypes (i.e and not through a clear relation of domination), and how dancers can refuse to occupy particular subject positions which place them in uneasy positions of coherence. In a sense these resistances are, indeed, effects. An alternative approach would be to replace these effects with a subject capable of resistance, but this runs the risk of essentializing the subject and 'romanticizing' the concept of resistance (Abu-Lughod 1990). To focus instead on how individual resistances reveal power in particular contexts and how these resistances, taken together, play a role in defining the contested terrain of politics in/at the bar may be another way to conceptualize resistance while simultaneously displacing dichotomized relations of power. Within these domains there are mobile and multiple points of resistance which surface in dancers' relationships with customers, researchers, management and between themselves, and at issue are more personalized questions of self-actualization, personal ethics, friendship, morality, and so on.

One evening in the Brunswick, and this approach to resistance raises interesting questions about what assumptions of power, agency and encounters between people have been assumed in traditional analyses of resistance. In reconceptualizing the space of the bar as a negotiated space, a third space where identities are negotiated and ambivalent, performed and not fixed, it becomes possible to question the positioning of dancers as 'victims'. While the 'reality' of sex tourism in the Philippines is a historically specific reality, the practice of sex tourism, and the various identities of the bar are certainly more ambiguous. Understanding the manner in which points of resistance surface in the contested terrain of the bar offers a way to conceptualize the politics of meaning within the sites of sex tourism, as well as our comprehension of geographies of resistance.

4

BEYOND THE BAR

Lives, community and transient identities

> You can take the girl out of the bar, but you can't take the bar out of the girl.
>
> (famous cliché about Southest Asian 'bar girls')

Very little of what had been written about prostitution in Southeast Asia prepared me for the chaos of what social theorists would call the 'everyday life' of women in the sex industry. Throughout the course of my extended ethnographic fieldwork in the Philippines (1992–94), my well-rehearsed arguments about the political economy of prostitution became an increasingly abstract construction of academic and activist discourse and – while providing insight into the problems inherited from colonialism, militarism and sex tourism – played a rather ambivalent role in my understanding of the complex subjectivities of the women that I knew. The result of my 'ethnographic anxiety' was twofold.[1] First, I resisted these rationalizing discourses through articulating, and encouraging discussion on, their inconsistencies and normalization strategies with both dancers and NGOs. Later, I withdrew from the NGO scene almost entirely and absorbed myself in ethnographic research – unable to deal with what appeared to be the incommensurability of these two worlds. Anyone who visited or received letters from me during this time was well aware that I had developed an attitude of 'leave dancers alone – you're making it worse'.

At the time I considered the possibility of narcissism in my research; in other words, I worried about my own arrogance in suggesting that middle-class NGO workers did not understand these women as I did. This feeling was exacerbated on occasions when I declined offers to participate in NGO anti-prostitution projects (i.e. because it contradicted my ethnographic work). Through the process of reading, writing and ordering the chaos of my field notes, however, I now realize that I had been resisting the inevitability of a partial perspective – including mine. Therefore, my resistance to the simplicity of modes of theorizing on prostitution can also be understood as a response to my own failed attempts at constructing a coherent prostitute 'subject'. I was resisting the NGO construction because of its inconsistencies, there were too many silences/absences, and the edges were left completely unexplored – but I could offer nothing 'substantial' in its place. Furthermore, I had observed that these simplified constructions eventually intervened in women's lives through the course of development/education projects, where sex workers learned to perform the victim identity prescribed by the project. What I had yet to learn was how donors and NGOs drew on these simplified constructions as part of a repertoire of rules/resources in making sense of, and gaining control over, their own lives and professional situations (see Chapter 6).

In this chapter I aim to allow for the complexities, paradoxes and richness of sex worker subjectivities, while at the same time giving meaning to my own oppositional tendencies in the first instance, and to my ultimate withdrawal from discussions in the NGO community. In this sense I also aim to destabilize the boundary between theory and empirics, where my experience *becomes* theory.[2] My difficulty in constructing a homogeneous prostitute subject was not a result of my inadequacy as a social researcher – although I did suffer this sense on a number of occasions – but a response to the complex ways in which my 'research subjects' and I made sense of ourselves, of each other and the rest of the world. Contemplating the subject of prostitution without interrogating my own conceptualization of sex work, without understanding why Cora thought I was a religious or social worker or why bar women's perceptions of their work changed on a regular basis, became unthinkable. These insights, obtained by exploring the boundaries of my research, appeared virtually absent in the discourse of NGOs.

What follows can only be a slice of that process, and I have strategically chosen to elaborate the ambivalence of sex workers' narratives about their entry into, and employment in, the sex industry. These stories give women a place to think, feel and act, recognizing their position and place in the world. These narratives explore the differing constructions of the meaning of bar work and personal identity, and highlight the inherent multiplicity within the subject herself. Following Kondo's (1990: 44) conception of the self, this chapter therefore examines selves as 'potential sites for the play of multiple discourses and shifting, multiple subject-positions'. The emphasis here, however, will be on how these shifting and multiple subject positions intersect their lives/bodies, as well as how these intersections are linked to spaces such as the bar, the neighbourhood, past residences, and their sites of lost virginity. This perspective allows them to be involved in something inside, outside and *beyond* maps of meaning and power, and to be recognized as mapping the worlds they live in.

By allowing sex workers to speak about their employment in the bars of Fuente Osmeña, what becomes evident is that these women do not author themselves as homogeneous victims of colonialism, sex tourism or political economy. Prostitution is only one facet of their lives; it is a job they are employed in – often for short periods of time – and cannot be equated with the essence of their personhood. Bar women have split subject positions and multiple identities: just as they ambiguously construct themselves not just as prostitutes but also as interpreters, local guides and prospective wives *inside* the bar, these women are also mothers, daughters, wives, girlfriends and income earners *outside* the bar. But these identities are not necessarily separate personas: some overlap and are compatible, others are transient and incompatible.

Acknowledging this complexity is *contra* the simplifications offered by NGOs, and in a more quotidian sense, *contra* the opening quote to this chapter. 'You can take the girl out of the bar, but you can't take the bar out of the girl' is a famous cliché in the bars of Cebu, and has been proclaimed by many customers/boyfriends who have watched their girlfriends enter and exit the industry on a regular basis. That many women take offence to this evaluation of their personhood is hardly surprizing; the cliché is usually offered in a negative way and infers promiscuity beyond the bounds of an imagined Filipina womanhood. Yet I rarely, if ever, met a sex worker who would oppose alternative employment if something financially comparable were available, and most women are adamant that changing their presumably 'promiscuous' life merely requires changing their employment. Such responses to this cliché are, I think, emblematic of the result of containing identity from the 'outside' – of denying the complexity of their lives. Which is not to say, however, that this cliché cannot be creatively appropriated to end a relationship, or to justify particular kinds of personal behaviour.

The ways in which bar women struggle over issues of identity and bar work are neglected in the discourses of prostitution in the Philippines. To address this absence, I recount the stories of three women who have worked as dancers in the bars of Fuente Osmeña.[3] This is not to reveal the differences of their individual subject positions – although their differences are important – but rather to suggest how multiple subjects inhere within each self. These multiple subjects are then examined in terms of their relationship to the spaces they inhabit. Before telling these stories, however, I want to address the ethnographic issues in writing about these three women.

Writing (about) three women

> Ethnography is actively situated *between* powerful systems of meaning. It poses its questions at the boundaries of civilisations, cultures, classes, races, and genders. Ethnography decodes and recodes, telling the grounds of collective order and diversity, inclusion and exclusion. It describes processes of innovation and structuration, and is itself part of these processes.
>
> (Clifford 1986: 2–3)

Politically and epistemologically, the project of ethnographic description is contested domain. As the proverbial 'crisis of representation' echoes the reverie of post-structuralist thought, it concedes the impossibility of representing 'others' except through our own interpretive schema. This crisis in ethnography, far from merely celebrating the death of the author, rather accents the important issues of power, authority and representation in the production of ethnographic texts. To address these issues in my accounts of three Filipina dancers, I draw on insights gained from writing in reflexive anthropology (Clifford and Marcus 1986; Marcus and Fischer 1986; Clifford 1988), including Clifford's (1986) essay on 'partial truths'.[4]

Clifford makes two points worth rehearsing here, particularly because of their affinity with Bhabha's (1990, 1994) notions of between-ness, hybridity and third space. First, in the introductory quote to this section, Clifford situates ethnography '*between* powerful systems of meaning.' In general terms, he is emphasizing that the world of the ethnographer and the world of the researched have different structures of meaning – this much is obvious. As a consequence, however, certain voices and certain texts will have more or less established meanings in the encounter between the ethnographer and the researched, and therefore be included and/or excluded in the final (coherent) ethnographic product. In the case of my own accounts, I attempt to make explicit when, where and why inclusions are taking place; exclusions, although anticipated, are for the most part determined by the audience.[5] In emphasizing my own experience as theory, for example – by realizing that the stories of Filipino women were told and listened to for specific reasons – I do not essentialize but question and point beyond the identity of dancers; in this sense I create something between the lived experience of sex work and my own academic interests. The specific life histories described below are not about a more detailed representation of prostitute subjectivity, nor are they about the post-structural turn in the social sciences; both, however, have greatly influenced what is described.

In discussing the limitations of 'reflexive anthropology', geographers such as Keith (1992: 554) have argued that just as representation (necessarily) misrepresents, ethnography (necessarily) betrays: ethnography is always

> an act of betrayal . . . individual researchers [are] never invisible, they are also always malleable, and hence inevitably duplicitous, figures in the background of the research process . . . [which] in the final analysis will always give the academic author the final word.

Keith's predisposition to a progressive politics disparages an emphasis on exclusively aesthetic and/or textual ethnography. While I would argue that reflexive anthropology has been instrumental in questioning authority, the closure of texts, and has pushed out the boundaries of what the 'social' might look like, I take Keith's point seriously. I therefore address questions about how representative these stories are in terms of their origins, itineraries and trajectories (e.g. poverty, education, familial support, etc.), as well as examining the broader commonalities that might reveal a space for a speaking prostitute subject. This space – which I refer to as the third space of narrative – is then reconsidered in the realm of politics in the following chapters on government and NGO interventions.

Second, Clifford (1986: 26) maintains that 'ethnography is a hybrid textual activity' which 'traverses genres and disciplines'. While Clifford is discussing the dangers of what might be considered excessively 'self-conscious' and 'poetic' ethnographic texts – dangers that Keith (1992) contends may form an impediment to politics – he also stresses that ethnography is always about writing. 'Writing culture' necessarily draws on a diverse array of field experiences, theoretical perspectives and literary methods, and this hybrid conception of ethnography has much affinity with Bhabha's notions of hybridity and third space, and acknowledges the product of ethnographic inquiry: texts. Just as Bhabha performs an ambiguous postcolonial subject position in his writing (Rose 1995), I attempt to write through and beyond, and not necessarily about, working in Cebu's sex industry. I am therefore defining a third space for both myself and Cebu's dancers by *writing (about) three women*.

Virgie: My son has made me very strong

'I don't have enough education, I have no choice. It's easy money. I have no family here . . . ' – Virgie uttered these statements in response to my standard questions regarding her employment in the sex industry.[6] We were sitting in the main living room of her two-bedroom house this afternoon, both on our best behaviour. Although I had been to her house on other occasions, it was evident that this time we both felt under scrutiny. I wanted more detailed information on her employment for my research – Virgie's articulate stories in particular played a role in subverting the 'victim' discourse espoused by NGOs – and she was considering the prospect of my renting her extra bedroom. She was worried about how I would adjust to living with her son Joey who, at the age of three, was the biggest bundle of energy I had known. I said his hyperactivity was not a bother to me, and after he'd furtively drank most of Virgie's glass of beer and went to sleep, she told me the following story of how she came to be employed in the Brunswick.

Virgie was born in rural Negros and is the eldest of eight children. She maintained residence in Negros until she completed high school and a visitor from Cagayan de Oro, Mindanao suggested she migrate to Cagayan to seek employment. Exhausted by the duties and responsibilities of the eldest daughter, she agreed, and worked in Cagayan as a domestic helper for several months. The death of her mother caused her to return briefly to Negros, but after the funeral her father was disappointed with her decision to continue employment in Cagayan. Being the oldest daughter Virgie 'should' have resumed residence in Negros and taken over the responsibilities of her mother, but the taste of city life was too appealing. This disagreement caused Virgie to lose regular contact with the family: 'From that day I knew I was really on my own,' Virgie said, 'and that I would have to work very hard to get the respect of my father.'

In Cagayan Virgie befriended another domestic helper and together they decided to migrate to the growing city of Cebu. She had passed through Cebu on her way to her mother's funeral and had met a friendly woman there who helped her transfer boats. Virgie was depressed and disoriented at the time, and convinced this woman was sent from God – she decided that this was a sign that her luck would be better in Cebu. After a few months of low-paying jobs in Cebu's restaurants, however, Virgie resorted to the classified advertisements in the local newspaper. One was for a waitress in a restaurant for tourists in Boracay, with a salary of P2000 (approximately US$80) per month.[7] She took a 20-hour boat ride to the island, but soon after arriving learned that her salary would only be P700 per month. Virgie and her co-workers were earning small salaries for harsh labour, and sometimes vented their frustration by attending a disco after work. It was here that Virgie met the Filipino father of her son:

> But one night we decided we're going to leave this job. What I remember is that we were throwing our things down from the upstairs and jumping down after them. We went to my friend's house and they call Ed for me. He took us to his house and I lost my virginity. But his family is rich and I didn't feel comfortable, so we say we'll meet in Cebu. But I never saw him again.

Upon returning to Cebu she was unemployed and realized she was pregnant. A former employer had opened a small restaurant, and she worked there until she was ready to deliver her son. After the birth, she was in desperate need of money and began work as a security guard. She'd been working for one day when she passed by the Red Bar and noticed the sign on the door that said 'receptionists wanted'. She inquired, was hired and agreed to start immediately.

Her first night she was bought five 'ladies drinks', so she earned a P30 salary and P100 in commissions on drinks that night (US$5.20). This relatively large amount of money she spent on milk and other things for her son. She met an American on her third night, they 'fell in love', she stopped working and he supported her for five months. She's unsure as to why he discontinued the relationship, but thinks it may have been sabotage by her jealous house-mate.

> Working in a bar is OK because I have self-respect. Some women are in despair thinking that no-one will love them any more. They

> drink, take drugs and see no problem with what they're doing because, anyway, they have no future. I'm always positive in my thinking, I'm praying and saying 'help yourself'. Sometimes a lot of people are helping you and you don't know how to help yourself.

Eventually Virgie's savings ran out and she applied to work at the Brunswick. She was a dancer there for several months and cultivated emotional/sexual relationships with two regular tourists, one American, the other Australian. The American sent her money on a regular basis, and with this money she rented and furnished a house she felt was suitable for her son. The Australian, on the other hand, did not give her money until she was forced to resign because of a hastily introduced bar policy which prohibited the use of stockings while dancing. There were many perspectives on why this policy was introduced, but it should be noted that only two women wore stockings at that time. Virgie suspected that Fely, the manager of the bar, was jealous of her relationships; May claimed it was because Virgie was too quick in soliciting drinks and thus frustrating Fely's business efforts; and Fely claimed it was because customers were complaining about not being able to see women's bare legs. The policy was, nonetheless, a good excuse for Virgie to take a break from dancing. Virgie's resignation was submitted two weeks before the day of my interview, and she had decided that she was finished with bar work, and that the Australian would be her future husband. She wanted me to rent her extra room in order to supplement her income and to 'learn the ways of Australia'. Such was my entry into Virgie's life, and her into mine, or at least this was how I understood it at the time. It would be a mutual learning experience.

Filipina/Australian couples are a common phenomenon. In 1989–90 alone, 729 Filipina wives and 893 Filipina fiancées were sponsored to enter Australia (Smith and Kaminskas 1992).[8] There is an extensive literature on Filipina mail-order brides, but as in this case, many Australian men in the Philippines find marriage partners while travelling or working in various capacities in the country. Success stories of women who married customers and are now overseas are common, and it is therefore not surprizing that Virgie considered her Australian boyfriend a prospective husband.

When I asked Virgie, several months later, if she thought dancers would transfer to another job if given to opportunity, she said:

> Why should she go back to the hard life? She's already been there, that's why she's in the bar. Why be a martyr? Working in a bar is OK if you work hard. It makes you smarter. If a woman has low education, like grade 3 or 4, then if she works in a bar she becomes smarter. She gets to meet professional men, go to expensive restaurants, more than even a teacher can. She has more

> chances to experience and learn. Me, I'm just like this but I've met the owners of the big hotels here and we've been talking. Sometimes I say to myself, if I was staying in my province, I would not meet the owner of the Graduate Hotel. He knows where I come from . . . My friend, she's only grade 4, but she learned English and became more confident. She got a new hairstyle, some new clothes and people started calling her ma'am. She dresses in white and goes to the casino. When people ask her what she's doing on the jeepney, she says she's working the night shift at Chong Hua Hospital.

Virgie maintains that life as the oldest sister in rural Philippines is difficult, and worth getting away from if you have ambition. When Virgie says she has 'no choice' but to work in a bar, she relates this to lacking a college education – which is increasingly necessary even for unskilled jobs – but more so to needing money to raise her son. Without extended family nor an adequate welfare state, her son's future, and ultimately her own, were uncertain. She looks at her entry into the sex industry as being difficult, but as providing her with an opportunity to experience more than she would have experienced had she stayed at home. Furthermore, through her work she had the opportunity to meet men and 'fall in love'. Sex work in this form is seen as more than a job, and in some ways resembles what Hochschild (1983) describes as 'selling personality' and 'emotional labour'.

The Australian supported Virgie for the time I rented her extra bedroom, but the money he sent was erratic. Between my rent and his occasional gifts, Virgie was barely making ends meet. Virgie had a keen business sense and asked if I could loan her the money to run a small food stall. She ran this business for about two months, but on most days the business barely made a profit. When I left for Australia she asked if I would mind if she closed the stall even though she could not repay me, and I agreed it was a good idea. There seemed no point in running the stall apart from merely giving her something to do.

A few months after my departure to Australia, Virgie's boyfriend Nick helped her move back to Negros where she introduced him as her fiancé. Her new social status allowed her to made amends with her father, but this status was only short-lived. Within a few months the Australian ended their relationship, and in order to cope with her new responsibilities – she had enrolled her two younger sisters in high school – Virgie was faced with working in the sex industry again. The letters I have received from her since this time were initially depressed and overwhelmed, but have more recently become positive or optimistic.

> September 9/94: . . . it's too hurt my feelings . . . you know I have a practical mind but I can't deny that I loved Nick very much . . .

> until now I'm not feeling so well, feeling so bad and I'm not happy working in the Side Bar because mostly the girls there are hypocrites.
>
> October 20/94: . . . now I work so hard for my money, you know what I want for my life and Joey. Nick was saying to me in his recent letter 'don't worry about the future'. I think I'm not abnormal thinking about my future. He's so bullshit.
>
> November 16/94: . . . until now it's still painful to remember about the past, di ba?[9] But right now I feel very fine. I've met someone from Switzerland who is handsome, the same age as me, and is a kung fu student with no vices. You know before I like the old man because he had a lot of experience. But now when I observe a man who has a lot of experience, I observe more bullshit, like Nick.
>
> December 29/94: . . . Derek, a friend of Mr Nick, was here in Cebu City before Christmas day and he was here three times in the Side Bar. I didn't talk to him because I was feeling bad when I saw him. You know on his last night he came to me and hugged me very much. He said to me, 'Virgie I'm so sorry', and explaining to me all about how Nick's not really shit and that he still loves me very much. Do you believe him, Lis? They are both assholes. Derek asked me to write a letter to Nick who's sick and in the hospital for about one month. Derek is saying to me that Nick's new girlfriend is crazy, and I answer him, 'I don't care, it's not my business. I'm happy with my two boyfriends' . . . Lisa, my two boyfriends, they are both nice and have a different attitude. I feel I can enjoy them but I don't expect them in the future. Now I don't expect too much because I don't want to hurt again, di ba? . . . I'm so busy now with my job, writing penpals, making boyfriends – I'm just joking!
>
> June 24/95: I'm really in a very hard life now because all I earn I spend. My money is for Joey's yaya [nanny], plus my two sisters in high school, and everything is my responsibility. Oh what a life of a single mother and oldest sister in all. I hope next year changes my life for the better.

Virgie's story illustrates the shifting character of dancers' perceptions of their situation: the construction of bar work is dependent on the location from which it is espoused. While Virgie was unemployed and considering the prospect of marriage, bar work was the stepping stone which had brought her to the present moment, something that had given her opportunities and made her smarter and stronger. Once her employment resumed, however, bar work became the usual set of ups and downs.

May: It's the bar that needs me, not me who needs the bar

May and I first met on my first trip to the Philippines. I was conducting interviews in the City Health Department and she came over to tell me it was a useless task because no one would talk to me in such a sterile environment. Instead, she invited me to join her at the local café, where we drank coffee, smoked cigarettes, and she told me about her experiences living overseas. This conversation at the café symbolized the beginning of my abandonment of objective interviews, and the beginning of our friendship. May's confidence and attitude intrigued me. As my fieldwork progressed, May took on the role of my friend and informant, often explaining the details of bar life she thought Virgie might be silencing. After several requests for a formal and taped interview, May invited me to her house. Unlike many of the dancers I met, she was not ashamed to extend an invitation. Her neighbours were fully aware of her employment: May wanted to live her life this way, and argued that if her neighbours disagreed with her choices in life, they were not worth having as friends. On this day, May told me the following story.

May was born in 1959 in a poor *barangay* of Cebu City. Because her mother left the family in the mid-1960s – she married an American serviceman who was stationed in Cebu at the time of the Vietnam War – May was raised by her father, and her father's sister:

> The side of my father, that auntie I have, I really hate her because I'm not supposed to be like this now, you know, if not for her. Because when I was a kid she really treated me bad . . . I've lived on my own since I was 13. Until now, I am 34, I am all by myself unless I have somebody like this [she motions to her sleeping boyfriend]. You know, he don't have a job.

When May was 13 years old she rebelled against her family by going to Bacolod, Negros. At this time, it was a long boat ride – more than 12 hours – to travel to this neighbouring island.

> And I was very lucky that I met some guy there then. A very old guy, he got 3 kids, they're girls also. And I am the fourth one, you know? Until I was 17 I stayed with them. So when I come back here . . . when my father and my auntie saw me there in the port, they called the police and then they take me home. And I remember that they take me to the hospital . . . to see if I am

> still a virgin. But I'm still a virgin at that time. So when I can, I get away from them again. I went to Zamboanga . . . I was just looking for a job there, but not this kind of job. And what I remember before, I hate 19. That's when I lose my virginity. And after that I worked because all my friends are saying 'come on you know how to dance, there is a club there that just opened'. I said no because . . . I never say that I am beautiful, OK? So I said, 'No, maybe they don't want me'. They said, 'No you can do it'. So I said, 'OK, let me try'. So that's where I stayed then, and that's my first job in Zamboanga.

In 1979 May returned to Cebu to find that her father had died, and she began work at the Side Bar. After one year she met a German, went to Germany with him, and stayed there for seven years:

> Then in 1980 I went to Germany, I met this guy in the Side Bar and he wanted to take me. I said 'why not?' I went with him, and when we got there we stayed for six months together before he let me know that I wasn't allowed to stay for six months . . . I did not find out until I asked my girlfriend about my visa . . . And then after that I asked my boyfriend if he's going to marry me . . . and he said no he cannot yet because we are just staying together for a while, not a long time . . . And then I said, 'What am I doing here? You have to let me go.' And he don't want me to go home, so that's why he . . . found me a husband. And it took about two years before we get married . . . all my paperwork is in the Philippines . . . And then I could not get along with him . . . before our first anniversary I asked him for a divorce . . . and it takes him about three years before he decides he wants a divorce.

May returned to Cebu in 1987 and started a small food business at her house. In 1988, realizing that she could not support the lifestyle she was accustomed to in Germany, May began work at the Brunswick. It was here she met an Englishman who enticed her to Abu Dhabi on a fiancée visa.

> I went there [Abu Dhabi] in 1992. I was only there for six weeks. I did not know that we were not compatible, because he was very nice to me when he was there. And also, while I was here, I can get his money, because he sent it to me. But then when I got there I don't get any . . . That's the problem, because if you're inside the house – and I got a problem about that – but if he's outside talking to his friends, he get everything you know.

After six weeks May returned to Cebu and resumed work at the bar, and has been working on and off since:

> I'm not looking to my future. I've tried my best already before but it didn't work . . . As soon as possible, if I can make money without it, I will stop. What happens tomorrow I will take it. Some people say you have to take this money, it's for tomorrow. No. Same with my boyfriend. You cannot keep everything. Some people ask 'why don't you have savings, maybe you'll get sick and won't have it?' I will take it.

Few employees are allowed to work on again, off again without repercussions, but May's philosophy – 'it's the bar that needs me, not me who needs the bar' – allows her great freedom. Indeed, the reason she consented to one of my interviews was to have an alibi for a clandestine exchange with a customer before working hours. May sees the bar as convenient for her profession, which is nothing more than paid sex. May's outlook causes some tension among the women in the bar, and those who are striving for a 'better life' (i.e. a husband) almost resent May's philosophy. As Virgie once said:

> Talk to May, she's been married once and then engaged. We ask her, 'May, why do you always keep coming back to this place?' Then May says, 'My first husband, we have a nice family, but I don't know. He's so . . . '. May, she's always going out bar hopping in Germany, but her husband doesn't like. So she's not happy and came back to the Philippines. She gets engaged to an Englishman and moved to Saudi Arabia. The same problems and she came back. We ask her, 'Why you not ashamed, May?' But she doesn't care. He didn't give her money, she's always staying at home. Because if you're a wife it's different, you're not going out any more. But May's like a single woman . . . We ask her what about when you're 50, May, you can't dance any more. May says she's already had a happy life, even if she dies tomorrow. She's not helping herself . . . there's no future if you're always dancing.

As discussed in Chapter 3, May's entry into the sex industry is a common one. The contradictions of male desire, and May's experiences in Germany and Abu Dhabi taught her that life in a bar allows you more freedom than being confined – as a Filipina wife can be – to the house. But Virgie's perspective also highlights the differing realities of bar work amongst the women. May and Virgie, for example, were never friends, they never socialized together, and I often felt caught between their two very different renderings of bar work.

When I returned for a brief visit to Cebu in mid-1994, May was working part-time in the City Health Department as part of an ongoing education programme with Cebu's sex industry workers. She said she enjoyed the work, wearing the uniform, and so on, but had been experiencing difficulties because sometimes, due to late barfines, she could not make it into work on time. I asked if she would read my rendering of her life history, which I had constructed in her absence, and give me any comments that might be helpful. May said she thought it was an accurate portrayal but that I'd missed the most important part. She's said I'd forgotten to mention that meeting her mother was the only dream she hadn't had yet. Later I went back to my notes and found the following quote I feel obliged to transcribe. In the interview, May was relating her difficulties in Germany, not knowing what to do about her visa, whether she was going to be arrested, and generally feeling alone. She telephoned her mother in the United States, and asked her for help. Her mother resisted because her new family did not know she had a daughter.

> And so that's why I said to her on the phone, you know, 'OK, never mind. Anyway people die, and I know you're going back to the Philippines so don't tell me that you buried yourself there. I know you're going back and I will spit on your grave.' I told her that. 'Because you don't want me, OK.' That's why when her mother died, in 1988, I was sad and happy because I know she's coming. Of course that's her mother. But she did not because she knew that I was around. So that's why I told my family, 'I don't know what I'm going to do with her but just let me see her face.' That's the only dream that I haven't had yet. That's why I said, Lisa, that I have everything already before, and I have done this and that before already. And I think I cannot ask Him more. What I have now is just like that. I'm not preparing myself for future any more. The only dream that I never had, that I'm asking Him, and that he never gave me yet. Even if I die the same day that I see her face . . . I really want to see her but I still hate her because of what she did. But I still want to see her just to look, I don't have to talk . . . Because I know that as soon as she starts talking to me I will feel something, because of course, that's my mother. And maybe I will feel something good, or maybe I will feel something bad, and that's why I'm afraid . . . Because anyway she doesn't feel like a mother to me, because she was never there when I grew up. Maybe what I'll feel towards her, if I see her, is maybe like a new friend, or friend before that I've never met yet, just meet like that. I don't know.

I had initially resisted using this quote because it was in danger of supporting the notion that all dancers come from dysfunctional families.

I feel more confident including it here now, because I realize that May was able to tell me this story because I had similarly lost contact with my own mother. Neither May nor I had spoken to our mothers in several years, an absence we both felt played an important role in who we were.

Edna: My mother didn't want us to marry a Filipino who doesn't work

> It's hard working in a bar. It's hard to go with customers especially if he's a sadist and wants to see blood first. It's hard if he's a married man. I hope I find a man so I can stop working. Sometimes I feel bad for two or three days. I feel bad about my job, I'm always thinking. But everywhere you need a college education. For me I'd like to work in a department store and not go out with customers, but I have only grade 6 [schooling]. It's a nice life working in a store . . . It's not like America where it doesn't matter. My sister's a chambermaid at the Hyatt and she's only grade 4.

Edna told me this story one afternoon long after we met. We were in the kitchen of her house, where we always sat, but this time I had brought my pen and paper. The house is typically small and she lived there with her daughter, mother, sister and cousin. Edna had always resisted a 'formal' interview, but on this occasion she permitted me to ask questions which would require difficult answers. This was permissible because, and only because, we had reached a more-or-less comfortable level of friendship.

Edna's extended family can be traced to the sugarcane plantations of northern Negros. When she was 4 years old her parents migrated to Cebu to seek employment. She finished her education up to grade 6 and, as is common in her seaside neighbourhood, began employment producing shell craft on a piecework basis. She continued in the shell craft industry until she was 18, when she decided to join her sisters, all three working in bars catering for American servicemen in Angeles City. As she says, 'My mother didn't want us to marry a Filipino who doesn't work.'

I never had the opportunity to speak directly to Edna's mother. Whenever I arrived at the house she would grab my hand and hold it deeply. I knew she'd had a difficult life, but this information came primarily from her sister Margaret, a woman I usually referred to as Manang.[10] Manang lived in the same neighbourhood and I visited her on many occasions because she practised traditional massage and was a link to the Negros folktales that played an important role in my everyday life with Virgie and her son. Virgie and I visited Edna's community together on a regular basis, and Virgie and Edna were very good friends. Manang's daughter also worked in a Fuente Osmeña bar, but for the most part, refused to talk to me. Manang frequently told stories about the horrors of World War II and Japanese occupation in Negros – their family home was burned to the ground – and the joy of American liberation. Two of her daughters had worked at the American military bases as dancers, and both married American servicemen. She was delighted, and hoped that her daughter Linda would do the same, but Linda was married to a Filipino who did not work and, in much the same way as May's boyfriend, lived on Linda's bar money.

Many studies document the history and ill effects of military bases on women in the Philippines.[11] In general, the militarization of prostitution is seen as a complex relation of domination involving questions of national sovereignty, racism, class inequalities and sexual politics. Although the American bases at Angeles and Olongapo had their heyday during the Vietnam War, in the first quarter of 1990 there were a total of 1,567 registered entertainment establishments in Angeles and 615 in Olongapo, with a total of approximately 55,000 'hospitality women' in both cities (Santos 1992). The 'new world order' and the expiry of the US-RP Military Base Agreement spelled the end for the bases, but the Mount Pinatubo eruption in 1991 saw the unequivocal departure of American servicemen. This event also displaced thousands of dancers in both cities, many of whom came from communities such as Edna's.

Edna's two oldest sisters are currently in the United States, having married American servicemen they met in Angeles. Both sisters send remittances to Cebu, although often at irregular intervals. The third oldest sister also worked in Angeles, but feels unsuccessful since she did not find a husband. She now lives with Edna, taking care of her needs and corresponding with international pen pals in the hope of going overseas. Edna is the youngest and, apart from the remittances from the US, carries primary financial responsibility for the family. 'I get angry sometimes, but everyone understands,' she says, as her sister returns from a gambling session after losing the P25 that Edna had given her for cigarettes.

Edna was 23 when her two older sisters migrated to America, and she decided to work in a bar too. After four months a man asked if he could take her out for the night:

> My sisters had left and I had no job. So I danced for four months as a virgin. I was just strong because no-one can help me. Everyone left me . . . Then this man wants to take me out. They told me you go, it's only dinner and dancing. I didn't know at that time because I was innocent. He was a flier man in the military. He's forcing me. I'm screaming. The hotel doesn't care because he's a guest.

Edna was the only woman I met whose first barfine would be classified as a rape. This passage raises two interesting issues that I will only briefly mention, since a discussion of rape is beyond the scope of this chapter.

First, it is important to note that it was the rape, and not bar work, which contributed to Edna's loss of innocence. Second, the hotel is blamed for treating the American as a 'guest', as a paying customer, which presumably gave him a position of power. Edna's view of Filipino conceptions of justice was common in her neighbourhood. It was frequently intonated that Filipinos would do anything for 'dollars', even if it meant jeopardizing their relationships with other Filipinos.

Edna was 24 when she fell in love with an American serviceman in the bar. She resigned from her job and became pregnant. He wanted her to abort the child, but when she refused, he ended the relationship. They 'weren't compatible and [were] always arguing', she remembers, and 'he was butterfly, always going around with other girls in front of me'. She gave birth to a daughter and stopped working for two years. When Mount Pinatubo erupted the father's financial support ended, however, and she returned to Cebu with her daughter but without an income.

> When we were there in Angeles the Base made him pay me support. But then after Mount Pinatubo nobody is there to make him pay me . . . Now he's married to another Filipina and they're in the US. His mother told me that he don't have any daughter there, but he will not send money for his daughter here in Cebu.

Edna returned to Cebu because the physical destruction and expedited departure of the American servicemen decreased the demand for sex workers in the city. Liza, a friend from Cebu and Angeles, had taken up a job in the Brunswick and introduced Edna to the owner. She worked there for a year, but transferred to the Red Bar after having problems with the owner, who had wanted her to be a witness in a case against the disc jockey, cashier and bartender, all presumed guilty of taking money and sleeping on the job. Edna didn't want to be involved, however, so it was easier to transfer bars. She's been there ever since.

> I don't like to talk to customers. I'm ashamed even though I've been working in the bar a long time. Even the owner says I'm losing my salary . . . People look down on you. That's why I just don't listen to the *tsismis* [gossip]. I stay in the house and work for my daughter.

Filipino conceptions of shame play an important role in regulating behaviour – a conception intrinsically bound up with Catholic morality. Yet these women experience the 'shame' of immorality in various ways. Virgie portrays her life with rather Edenic qualities: a pure past (on the farm), a fall (her loss of virginity and subsequent pregnancy), and a struggle to survive (through dancing) which will redeem her in the future. May refuses

Filipino 'shame' through drawing on her learning experiences from Western countries, as well as with the customers she engages with in 'short time'. In Edna's case, shame plays a role in her relationships with customers and neighbours, and results in periods of depression that can last for several days. In all cases this sense of shame has spatial coordinates: they map life histories, and affect the way they move through the everyday space of the city and neighbourhood.

Edna is one of the many women who have returned to Cebu from Angeles and are working in Cebu's bars. As Virgie once said:

> People were saying that the Philippines could protect itself, but I don't think so. I was so angry at them, I remember all those people like Edna on the TV crying or throwing things because they were so angry . . . Some people they say Mount Pinatubo went up because of all the prostitutes there. I don't think so.

Although there are no published statistics on the number of women from Cebu who migrated to Angeles, Miralao *et al.* (1990) noted the increased number of women migrating to both Angeles and Olongapo from semiurban or urbanizing areas in the Visayas (e.g. Cebu, Negros, Leyte, Samar) during the late 1980s. This finding is augmented by Sturdevant and Stoltzfus (1992) and Santos (1992), and suggests that urban poverty is changing the profile of women in the sex industry.

Edna's seaside community was considered part of 'the province' in the 1960s and 1970s, but Cebu City's rapid growth has enveloped the area. It is a squatter community that has traditionally relied on the resources from the surrounding land and sea. The first settlers harvested *nipa* (thatch) for the construction of houses, the community then collected shells for export, but now that the shell resource is endangered from overuse and dynamite fishing, many are engaged in woodcraft. The supplies for woodcraft come from illegal logging activities, however, so the future of this activity is uncertain.

The community has an extensive history of women migrating to Angeles to work. The first 'bar girl' is rumoured to have migrated in the late 1960s, after which time a series of personal recruitments produced a steady flow of women to both Angeles and Olongapo. As Edna's own history indicates, many families rely on remittances from daughters working elsewhere in the Philippines. The number of 'Amerasian' children in her community, many with mothers that have Afro-American accents, is testimony to the liaisons established while American troops were stationed in the Philippines. Many families are now dependent on remittances from their children living in the US, however, and although this money is primarily used to keep up with the inflationary prices of goods and services in Cebu, some families have built foundations for their houses or started small businesses.

I spent much time documenting the life histories of several dancers living in Edna's neighbourhood, and the trajectories of their stories were similar. When I returned to Cebu in mid-1994, I offered Edna a copy of a paper I had written which included her 'story'. To the best of my knowledge she never read it, partly because it was in English, but said she was sure I did a good job since I understood the lives of dancers. Whether Edna's comment was to congratulate me or to combat what she perceived as my lack of confidence, I do not know. Edna taught me that constructing her as a 'dancer' was only possible through my academic pursuits.

Transient identities

As the stories of Virgie, May and Edna indicate, women's power and capacity to change their situation is clearly various. While Virgie affords herself the greatest degree of agency – she knows how to 'help herself' – she also cites the effects of oppressive family responsibilities, her loss of virginity and the limited employment opportunities for high school graduates as contributing to her entry into the sex industry ('I don't have enough education, I have no choice . . . I have no family here'). May also perceives her familial background as a site of oppression, suggesting that her entry into sex work is a product of being poorly brought up by her aunt ('I'm not supposed to be like this now if not for her'), but also of her loss of virginity ('I hate 19. That's when I lose my virginity. And after that I worked'). At 34 years of age, her unsuccessful relations with foreign men have contributed to her perception of bar work as sex work, a convenient occupation for a good lifestyle. Finally, Edna's family encouraged her to work in Angeles ('my mother didn't want us to marry a Filipino who doesn't work'), and while her loss of virginity did not play a role in her entering the sex industry, it contributed to her loss of 'innocence'. She continues in the sex industry despite feeling 'ashamed', however, due to a lack of alternatives ('I'd like to work in a department store . . . but I have only grade 6').

Furthermore, the women presented here have strong overseas contacts: Virgie plans for overseas migration, May has lived in Germany and Saudi Arabia and has a mother in the United States, and Edna's sisters and the father of her daughter live in the US. These women do not author themselves as part of a global sex traffic, however, largely because their work is perceived in relation to 'love' and relationships. Although foreigners are often nice in the Philippines and turn nasty overseas – when they become deceptive, stingy or incompatible – women perceive their 'foreign relations' in somewhat benign terms. They seem particularly benign when compared to interpretations offered by outside analysts in nationalist, anti-colonialist and/or feminist modes, as discussed in Chapter 6.

Most importantly, however, bar women do not perceive their identity as derived from, or determined by, their work; being a 'dancer' is a transient and flexible identity. Bar work is nevertheless a meaningful aspect of Virgie's, May's and Edna's lives, and the stories told here weave together dancers' shifting and multiple identities in relation to their entry into, and employment in, the sex industry. While my research interests undoubtedly played a role in this process – I have highlighted 'dancing' as opposed to 'mothering', for example – it is also important to stress that the role of 'shame' in Filipino culture plays a constitutive role in solidifying ambiguities into a coherent identity. Catholicism often invites women to embrace the status of the martyr or victim, and while dancers are able to credit themselves with worth and agency, the performance of a victim identity is significant. This performance is corporeal and has a geography: both Virgie and Edna, for example, wore exceptionally conservative clothing while in their neighbourhoods (as opposed to their sexy clothing in the bar), and spent most of their days in the house (because in the evenings they were drinking and dancing in the bar). If such performance is understood as an effect of dominant codes of morality rather than an essence of identity, however, it is possible to see how different spaces in the city offer different possible identities. In this way, an essentialized prostitute/sex worker subject can be challenged.

These aspects of dancers' lives are rarely discussed by NGOs. This is partly because the ambivalence, ambiguities and paradoxes of working in Cebu's sex industry are difficult to convey in a structural analysis that casts them as victims of political economy. Although NGOs also utilize the concept of a 'life history', this life is usually presented in terms of a specific political agenda: to highlight issues of foreign control, poverty or morality (a similar argument could be made for newspaper or magazine accounts, or television documentaries). On this basis, this chapter is no exception. A search for third space accounts acknowledges 'partiality' but is not relativist and holds the potential to imagine transient identities – an issue I will return to in Chapter 7.

Working in a bar is a meaningful facet of dancers' lives, and these meanings are shaped by the constraints of history, culture and politics/economics. The narratives presented here weave together these constraints in ways that create a new kind of politics of identity. The trajectories of these stories constitute, and are themselves constituted by, the dominant discourses of prostitution on the Philippines. Yet the narratives themselves are produced at the borders of everyday life. If campaigns against sex tours or the spread of HIV/AIDS are to be effective, then it is necessary to connect these everyday narratives with different, political narratives in ways which facilitate the social and cultural identification of dancers with NGO projects. This type of identification holds the possibility of finding new sites of collaboration between dancers and NGOs, and the following three chapters imagine how such identification might be developed.

5

SEX WORK, HIV/AIDS AND BLAME

Mandatory HIV antibody testing

As discussed in Chapter 2, the 'cartographies of desire' that map Southeast Asian sex industries have helped government and community organizations locate individuals vulnerable to HIV/AIDS. In many cases, and for reasons of data collection and surveillance, this has also meant that sex workers have been persuaded to have their blood tested for HIV antibodies. The contexts which surround testing vary from one country to another, but there are always tensions between perspectives which privilege the human rights of sex workers regarding their HIV status and those that foreground public health. Here I discuss the historical context of mandatory blood testing in Cebu City, contemporary debates about the rights of sex workers regarding testing policies and how this specific context has played a role in blaming sex workers for the spread of HIV/AIDS.

Studies of the effect of HIV antibody testing suggest that government testing programmes are not only expensive, but do not assure that people will change their sexual behaviour. On the one hand, there is a 'window period' for the incubation of the virus and a negative result does not necessarily mean that an individual is not infected. On the other hand, there is no guarantee that a positive test will result in practices of safe sex. Although HIV antibody testing has been criticized as giving out the wrong AIDS education message, it has been widely used throughout Southeast Asia, particularly for women in the sex industry. Testing mentalities have been not so much about reliability, however, than they have been about popular perceptions of prostitution and sex tourism.

Government intervention and blame

In 1992 an article on prostitution and AIDS appeared in a Philippine journal with the following introduction:

> The mounting number of reported AIDS cases in the Philippines cannot be neglected or ignored. This is largely due to hetero- and homosexual contact prevalent among hospitality girls, *especially in*

> *areas where foreigners abound*, such as Angeles and Subic where the military bases were [located] . . . [This] calls for *strict government intervention* in providing proper and effective channels through which information on AIDS and its transmission can be adequately disseminated.
>
> (Conaco 1992: 195, emphasis added)

Conaco's observation regarding the need for 'strict government intervention' in the lives of 'hospitality girls' highlights two inter-related themes: (1) the increasing role of public health authorities in official discussions of prostitution; and (2) and how anti-colonial sensibilities have inscribed the public health agenda. Because public health perspectives are authorized – and by this I mean the Department of Health professes a privileged access to reality – they shape public perceptions about HIV/AIDS. In the Philippines this has meant an emphasis on particular geographic areas, and the designation of sex workers as a 'high risk group'. This has not only reproduced stereotypes of sex workers as diseased, but has increased the stigma of working in the sex industry.

In 1993, there were debates about mandatory HIV antibody testing for women in Cebu City's sex industry. These debates were inspired by the Department of Health's national policy to eliminate mandatory blood screening on the grounds that it was expensive and ineffectual in slowing the spread of HIV (Chapter 2). While the rhetoric of the debate was influenced by the emerging 'global discourse' on HIV/AIDS and human rights, the debate itself was always about local perceptions of women in the sex industry and Cebu's administrative autonomy. Here I highlight the role of city politics in debates about mandatory testing (cf. Brown 1995), while at the same time reproducing an exclusionary discourse which silenced the voices of women in Cebu's sex industry. The voices which were represented in these debates – government and non-government organizations as well as the media – considered sex workers particularly vulnerable, but the debate itself exceeded discussions about preventing AIDS. Mandatory testing was simultaneously about the autonomy of Cebu's politicians, a history of regulating prostitution, and the global links that were perceived as bringing HIV to Cebu. In the process of creating meaning, however, women in the sex industry were scapegoated for the spread of HIV.

In order to address these issues and themes, I first give an introduction to HIV/AIDS in the Philippines, and how social and institutional perspectives have shaped perceptions of the epidemic since the mid-1980s. I argue that HIV/AIDS has been seen primarily as a virus of foreigners, and that sex workers – particularly, but not exclusively, those catering to foreign clientele – have been publicly implicated in its spread. Then I focus more specifically on Cebu City, the history of government regulation and the debate on mandatory blood testing. This debate emphasizes how politics are constituted in relation to AIDS prevention policies at national and global levels, and how the cultural politics of AIDS interweaved these scales in ways which resisted outside interference. Because this debate tended to reproduce the stereotype of prostitutes as diseased, however, it prevented the voices of Cebu's sex workers from entering the debate. This indicates that global strategies of anti-discrimination deployed by AIDS activists – while impacting on the local – can also be manipulated to reproduce stereotypes in contradictory ways.

HIV/AIDS in the Philippines

In 1984 the first Filipino national – a returning overseas contract worker – was diagnosed with AIDS. In the same year, eight prostitutes working around the American military bases at Angeles and Olongapo were diagnosed HIV positive. The subsequent designation of 'high risk groups' by both the media and government agencies led to almost a decade of portraying the spread of HIV in the Philippines as a problem of foreigners, and prostitutes became the target of much social hostility. This was most powerfully evidenced in Mayor Lim's 'clean-up' campaign which saw the closure of Manila's famous sex tourism industry, and forced dislocation of many women employed there.

Major participants in official discussions of AIDS in the Philippines include: the Catholic Church, the Department of Health (DOH), the World Health Organisation (WHO), and various international donor agencies (especially USAID and AusAID) and their recipient non-government organizations (NGOs).[1] The Catholic Church and its lay groups are internationally renowned, and are involved in what can be termed 'medico-moral' discourse (cf. Seidel 1993). They seek to control sexuality generally, condemn what are termed the 'three P's' – Premarital sex, Promiscuity and Prostitution – and as depicted in Figure 5.1, inspire great debate about whether or not condoms promote promiscuity and legitimize prostitution, or control the spread of HIV. The debate between the government and Church on AIDS has a longer history related to artificial birth control, but recent studies (Amante 1993; Dañguilan 1993) and the election of President Ramos, the first Protestant and pro-family planning head of state, suggest that the influence the Church once had over sexuality and contraception is decreasing.

The educational efforts of NGOs have been officially acknowledged as being crucial in slowing the spread of HIV (WHO 1989; Aggleton *et al.* 1993; Altman 1993), and in the past decade dozens of Filipino NGOs have become involved in HIV/AIDS education activities throughout the country. These NGOs, primarily community-based organizations, have reached groups otherwise neglected by government education programmes (the urban poor, sex workers, gay men, men who have sex with men), and have actively participated in discussions on national policy and human rights (Litong *et al.*

Figure 5.1 The Department of Health versus the Catholic Church position on condoms. This critique of the opposing stances of the Depatment of Health and the Catholic Church highlights an important debate in the Philippines. In this cartoon, Secretary of Health Flavier declares that condoms are 'protection for the AIDS crisis' while Cardinal Sin argues they 'promote sexual permissiveness'. While the two figures argue, AIDS creeps up in the background.

Source: *Philippine Daily Inquirer*, 2 February 1993

n.d.). While NGOs working with women in the sex industry have historically been feminist/nationalist organizations which highlight the adverse effects of an outward-looking economy on women, NGOs currently carrying out HIV/AIDS education projects are more likely to be health-based rather than feminist/nationalist. Indeed, health-based NGOs have been instrumental in shifting the neoMarxist perspective on prostitution to one that deals with personal empowerment to protect oneself from contracting HIV.

Due to prevention, control and surveillance functions, the WHO and DOH exercise micro-technologies of power which target 'high risk groups' and 'seropositives' – particularly among a group referred to as 'commercial sex workers' or 'CSWs'. Such a rigid construction of HIV/AIDS has led to the further control of women in the sex industry; not only must they undergo mandatory blood testing in government clinics, they are refused work permits if diagnosed positive. Furthermore, and often due to the nominal or minimal confidentiality afforded to sex workers regarding their HIV status, they are subject to much harassment, most recently evidenced in Ermita (e.g. police surveillance, media exposure). Although the DOH drafted a national policy in 1993 to eliminate mandatory testing on the grounds that it was discriminatory, expensive and ineffectual in slowing

Table 5.1 HIV/AIDS 'high risk groups', 1984–92

Group	*No. tested*	*No. positive*	*Rate/1000*
Men who have sex with men	2973	66	22
IV drug users	124	2	16
Military personnel	471	1	2
Female sex workers	176015	123	1
Overseas contract workers	520662	44	0
Blood donors	319410	18	0
Male sex workers	122	0	0
Male STD patients	267	0	0
Others	38537	112	3
Total	1058192	366	35

Source: AIDS Research Group, RITM (1993).

the spread of HIV, the devolution of health services allowed this decision to be made by local governments – an issue I will return to shortly.

Official DOH statistics on HIV/AIDS in the Philippines are presented in Table 5.1 and Figure 5.2. These statistics broadly suggest that 'high risk groups' have different infection rates, and that there have been a concentrated number of HIV cases detected around Manila and the US military bases. While this situation might have parallels with other countries in Asia, what I would like to emphasize here is the assumption that there are identifiable 'high risk groups' – of which female sex workers are one – and that the geography of the spread of HIV/AIDS can be traced to the areas around the National Capital Region and the American military bases.

Table 5.1 is a classic representation of infection according to 'high risk group' categories, and suggests that men who have sex with men and intravenous (IV) drug users have the highest rate of HIV infection, although not necessarily the highest number of HIV/AIDS cases. Female sex workers, overseas contract workers and blood donors, on the other hand, have a disproportionate number of HIV/AIDS cases, though low infection rates. While it is possible to interrogate these numbers on their own terms, there are problems with the representation of the data in this way. Not only does Table 5.1 assume that the categories are mutually exclusive, these statistics do not reveal a crucial social context: mandatory testing is required for female sex workers, overseas contract workers and blood donors, which directly influences their proportional representation, as well as their identification as a risk group in the first instance. Borrowing Cindy Patton's (1990) expression, these statistics could be understood as a process of 'inventing AIDS', where no representation of data is politically neutral.

Mandatory testing and the subsequent identification of HIV/AIDS cases among female sex workers led to their designation by the DOH as a 'high risk group'. This epidemiological designation and how such categorization encourages distancing, blame and stigma, have been documented

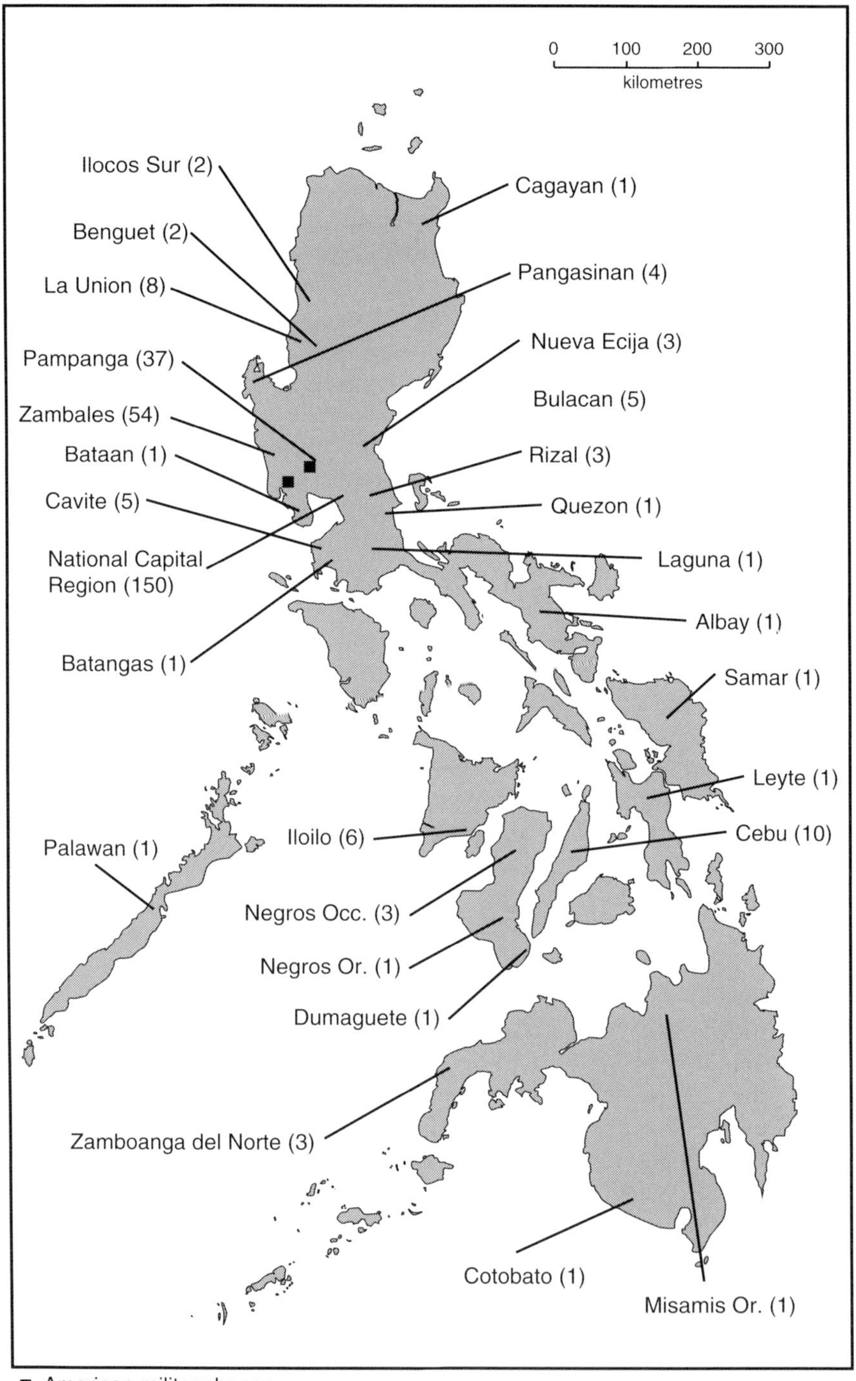

Figure 5.2 Geographical distribution of reported HIV/AIDS cases in the Philippines, 1984 to June 1993.

Source: AIDS Research Group, RITM (1993).

elsewhere – particularly in the West (Crimp 1989; Fee and Fox 1988; Kane and Mason 1992). In the Philippines, this has been most evident around Manila and the American military bases, where being female and engaged in sex work is often perceived as synonymous with being infected with HIV. It has also been prevalent in areas throughout the country where HIV positive sex workers have migrated to, once diagnosed positive. Although the DOH is said to conform to principles of confidentiality, for example, in a newspaper article discussing the migration of sex workers due to the closure of the bars in Ermita, the DOH warned that many of these women had migrated to Cebu and could be infected with HIV.[2] This particular incident excited discussion and local DOH intervention and was viewed as justifying mandatory blood screening.

Furthermore, and associated with naming and identifying high risk groups, is the search for the origin of HIV infection. This tendency was evident in the Soviet Union, where AIDS was initially conceived of as an American biological weapon, and in Europe where – 'along with jeans and rock' – AIDS was considered an American import (Gilman 1988: 264). The search has been most apparent in the United States, where the origin of AIDS has been alternately traced to Africa, homosexuals, IV drug users and ethnic minorities (Patton 1990; Gilman 1988), although in Japan there is evidence that Filipinos have been blamed for infecting the Japanese population (Buckley 1997). In the United States in particular, Gilman (1988: 266) suggests that 'the creation of the boundary between the infected and the healthy rested on the need to see a clear boundary existing between the heterosexual, non-IV drug using, white community and those at risk'. Some parallels can be drawn with the Philippines, although the boundary between self (healthy) and other (diseased) was appropriated for historically different reasons.

Figure 5.2 gives the geographical distribution of recorded HIV/AIDS cases in the Philippines. Although this representation is contingent upon the social construction of AIDS as a virus of foreigners, and sex workers as a high risk group, the map implicitly suggests that HIV/AIDS was brought into the country by men participating in the commercial sex industry in Manila and around the American military bases. While the virus may have been brought into the country through sexual contact with foreign nationals, these representations express more about popular conceptions about national sovereignty and development issues than HIV or AIDS. In short, HIV/AIDS is identified with foreign-dominated, outward-looking economic policies which encourage tourism – particularly sex tourism – and women in the sex industry are simultaneously cast as victims of political economy *and* vectors of transmission. This symbolic positioning characterizes these women as embodying an infectiousness that could affect the wider (healthy) population.

While blaming particular individuals or groups within society is a common response to a life-threatening virus such as HIV – whether the blame be placed on homosexuals, Africans, Americans, drug users or sex workers – the institutional and public emphasis on sex workers in the Philippines shares some continuity with historical representations of prostitution. This is not to say that sex workers are not at risk of becoming infected, it is the rather different assertion that conceptualizing sex workers as a high risk group cannot be separated from the conditions under which that construction is made (cf. Treichler 1989). Because prostitution has historically been constructed in relation to sex tourism by feminist/nationalist organizations, and because HIV/AIDS had been traced to outside sources, the labelling of sex workers as a high risk group appeared sensible. Such historical continuity is well illustrated through the situation in Cebu.

Regulating Cebu's sex industry

According to the City and Regional Health Departments, the first case of HIV/AIDS in Cebu City was diagnosed in 1989, and by September 1993 there were sixteen reported cases in the city. Press releases stated:[3]

> Of the 16 persons with HIV, two have died, two have been deported to their countries, two are undergoing treatment with private doctors, and three have not been located by monitoring DOH personnel. The seven are undergoing counselling. Many of them were commercial sex workers (CSWs) who remain the *captive group* to undergo *regular hygiene check* [*sic*], were between 20 to 45 years old, and were females. Only two of the 16 were male foreigners. Of the four cases recorded in 1993, two were males who are now undergoing treatment with private doctors and two were CSWs with one of them married.

Within this passage, the categorization and surveillance of these individuals are made quite explicit, as is the identification of 'CSWs' and 'foreigners' as high risk groups. My interest here, however, is the so-called 'captive group' of sex workers and their 'regular hygiene check'. Sex workers in Cebu have been required to submit to biannual blood screening for HIV since 1989, in addition to on-going weekly testing for sexually transmitted diseases (STDs).

During 1993, I collected information on government regulation of the sex industry in Cebu. This research was inspired by the institutionalized status of prostitution, where the greatest effect illegality appeared to have – at least for prostitutes registered with the city – was control by the City Health Department (i.e. mandatory health check-ups). This information was available in the Ordinances of the City of Cebu, but such ordinances dated only from 1945. There was little information about regulation before that time, although it is evident there was some form of regulation as early as 1932. In the 19 June 1932 editorial of *Progress*, for example, the editor dubs

Cebu the 'city of evils', and makes the following comments regarding the existence of cabarets and venereal disease in the city (Gwekoh 1932):

> By virtue of a 'wise municipal ordinance' dancing girls who, unless carefully inspected by competent health authorities, become the source of horrible venereal diseases, live and associate with the masses day and night without the least molestation by the authorities. For a handful of silver acquired in the form of taxes, government officials blindly lead the future pillars of the community to their physical and mental wreck and downfall.

Quite apart from its palpable morality, this passage represents some of the earlier attitudes towards the regulation of prostitution. Dancing girls are depicted as the 'source' of venereal diseases, yet government officials are represented as accommodating such activities for the benefit of public health and for financial gain. Public morals are juxtaposed with public health, in much the same way as they are today.

Although regulation in the 1930s was initially targeted at the flourishing industry of post-war cabarets, by the end of 1946, Section 3 of Ordinance 34 clarified the categories for, and extent of, regulation:

> Waitresses, hostesses and dancers in dancing halls, cabarets and nightclubs are required to submit to a bimonthly internal examination by the City Health Officer . . . and if found positive for venereal and other infectious diseases, their health certificates are temporarily revoked and they are temporarily suspended from engaging in their occupation. They shall submit to full treatment and confinement in the Social Hygiene Hospital until discharged negative after which they may resume work.

Regulation continued in much the same way until 1970, when the Sanitary Code of the City of Cebu (Ordinance 676) expanded the scope of the City Health Department's ability to control the spread of venereal diseases. Section 35.12, for example, stated:

> Whenever, on the judgement of the City Health Officer, the morbidity of venereal diseases undergoes a considerable and unaccustomed or abnormal increase, constituting a menace to the public health, said official shall have authority to adapt measures and regulations for the suppression or limiting of the infected prostitutes, and whenever possible, their isolation and treatment.

Although such ordinances have been revised over time, there are many continuities in the type of regulation. Women currently working in Cebu's sex industry, for example, must first pass a series of health examinations at the City Health Department's STD-AIDS Detection Center before they can obtain a 'hygiene card' which enables them to be employed as prostitutes. Upon commencement of sex work, women must return to the Center on a weekly basis for general STD testing, and since 1989 have undergone biannual blood screening for HIV. Although women are no longer confined to public hospitals, their health cards continue to be suspended if they are diagnosed positive with STDs. Treatment for certain STDs is offered within the Center, but for others, women must demonstrate that they have engaged in treatment by returning the packaging of their prescriptions to Center staff.

These Ordinances illustrate that although prostitution is illegal in Cebu, it has been controlled by local health officials since at least the 1930s. The social attitudes inscribed in the language of these ordinances over time are twofold: first, prostitutes have clearly been represented as the 'source' of STDs, and must therefore be strictly controlled and/or isolated; and second, under the Sanitary Code, the Department of Health has been given the authority to judge and adopt regulations to control the situation of public health within the city. These attitudes are implicit in Conaco's earlier quote, and are important to understanding the debate over mandatory HIV/AIDS screening in Cebu.

Random, voluntary and mandatory HIV/AIDS blood screening began in Cebu in 1986. In Table 5.2 – which I have reproduced from statistics provided by the Cebu City STD-AIDS Detection Center – the numbers of HIV/AIDS examinations conducted between 1986 and 1991 are listed according to categories defined by the Center. While these categories are not the traditional 'high risk groups' outlined above, this categorization suggests there are three essential groups – male, homosexual and female – which are popularly understood as: foreign male or male overseas contract worker; promiscuous homosexual; and female sex worker.

Cebu's first case of HIV/AIDS was diagnosed in a 'non-Filipino' male in 1989. Because sex workers had been identified as a 'high risk group' in other parts of the country, Cebu's sex industry came under close scrutiny and the City Health Department began mandatory HIV antibody testing in the same year. Blood screening did not reveal cases of HIV/AIDS amongst sex workers until 1991, and public concern increased in 1992 when reports in the local and national papers suggested there were many sex workers migrating to Cebu due to the eruption of Mount Pinatubo and the departure of the American military bases at Angeles and Olongapo.[4] Cebu's sex tourism industry was undergoing something of a 'revival' during the early 1990s, and this, combined with the semiotic links between Pinatubo and the dispersal of sex workers, combined to induce fear. Because of the concentration of HIV cases detected around these areas, these women were presumed to be infected and spreading the virus in Cebu.

Table 5.2 Number of HIV/AIDS examinations, 1986–91

	1986	*1987*	*1988*	*1989*	*1990*	*1991*
Males						
Filipino	6	52	1	11	3*	6
Non-Filipino	–	–	–	1**	–	6
Homosexual						
Filipino	35	26	39	–	10	1*
Non-Filipino	–	–	–	–	–	–
Females						
Filipino	996	643	733	791	826	2170**
Non-Filipino	–	–	–	–	–	–
Total	1037	721	773	802	839	2182

Source: Cebu City Health Department (1992).

Notes:
* Two diagnosed with HIV/AIDS.
** One diagnosed with HIV/AIDS.

Public concern intensified again in March 1993, when a foreign-owned experimental clinic for treating HIV and AIDS patients was discovered and raided by police. The clinic was using ozone therapy to treat seventeen foreign patients, primarily Americans and Australians, and was found to be operating without the appropriate licence or permit. The discovery of the clinic raised issues about local sovereignty and Philippine laws, particularly since the national government had recently discontinued the requirement of AIDS Clearances Certificates for foreigners visiting the Philippines. Although the AIDS clinic episode raised these legal issues, it also inspired the Mayor of Cebu City to suggest a 24-hour test centre for anyone who had contact with foreigners in the preceding weeks. He also assured Cebu's public that he would triple the efforts in having sex workers tested.[5]

The emphasis on foreigners and sex workers was explicit, but if read in a different context also reveals an ambivalence in reactions to Cebu's new economic status. Cebu has experienced rapid economic growth since the late 1980s, which has brought a large number of foreign investors and tourists to the city. Many NGOs in Cebu, particularly organizations against tourist development, have suggested that the outward-looking policies of Cebu's officials have been responsible for the deteriorating status of the urban poor, particularly poor women. While the Mayor has actively encouraged these global links, he has not welcomed the advent of AIDS. In his hurried response to the ozone clinic, however, and without concern for perceptions from the international business community, he reproduced the construction of HIV/AIDS as a problem of foreigners and sex workers.

By mid-1993, the City Health Department had identified, and begun surveillance on, a new group: injecting drug users. Anxiety over IV drug use initially surfaced in 1992, when 114 cases of cerebral malaria were detected and monitored by the DOH. By August 1993, another 69 cases had been documented. Because Cebu had otherwise been declared malaria-free, and because cerebral malaria had only been detected in urban areas, an inquiry into the issue was launched by the DOH. The inquiry revealed that 88.6 per cent of the cases in 1992 could be traced to injecting drug use, and more specifically, to injecting a prescription pain-killer called Nubain (nalbuphine hydrochloride).[6] There was much concern that the spread of cerebral malaria could be paralleled by HIV/AIDS, and government intervention was legitimized on this basis, together with suspicions of illegal transfers from Cebu's drug distributors and hospitals to the black market. This inspired national legislation to regulate the sale of Nubain, and the price of an ampule increased 700 per cent overnight.

The DOH produced figures that there were approximately 10,000 IV drug users in Cebu, the largest IV population in the country (DOH AIDS Unit 1993). Kamagayan, a neighbourhood in the Junquera district notorious for drugs and prostitution (Chapter 2), became the suspected Nubain 'shooting gallery' by both public health officials and police. Stories of Cebu's injecting drug scene – particularly the rumoured 'holes in the wall', where customers merely extended a cash-filled hand and arm into a small room for injection – reached as far as Manila. Because IV drug use was linked to the spread of HIV/AIDS, the National AIDS Prevention and Control Programme organized a consultative meeting on IV drug use and HIV in Cebu. The participants discussed issues such as the scope of the problem, research initiatives and intervention possibilities in Cebu. A well-publicized recommendation was to begin a community-based programme in Kamagayan, which would include needle exchanges, the provision of bleach to users and outreach and education. The programme was rejected by some local officials, however, who saw it as a locally insensitive WHO AIDS prevention agenda, with the possibility of increasing drug use within the city.[7]

As these events were transpiring, however, the DOH, in cooperation with the newly formed Philippine National AIDS Council, was drafting a national policy to eliminate mandatory testing on the grounds that it was expensive and ineffectual in slowing the spread of HIV. As in the case of banning mandatory testing for foreigners visiting the Philippines, this policy was inspired by global imperatives for more effective AIDS prevention. Many individuals in international organizations, aid agencies and NGOs had concluded a testing mentality encouraged stigma and blame, and was not an effective means to empower people to protect themselves from HIV. Because the devolution of health services presumably allowed this decision to be made at the local level, however, Cebu's politicians and bureaucrats argued against conforming to the national policy. The ozone clinic, the first cases of HIV in Cebu and the discovery of an IV drug-using population were all issues over which the local government sought to exercise its control. In each case, local officials felt best suited to assess the situation, yet were now subject to a Manila-focused national strategy that had been influenced by global imperatives. The ozone clinic, for example, prompted local officials to question the wisdom in declaring AIDS Clearances Certificates discriminatory, when their existence may have prevented the AIDS patients from entering the country (and presumably, infecting the local population). Furthermore, the suggestion of a needle exchange programme for IV drug users was seen to legitimize – not to mention organize those affected by – economic and social problems in Kamagayan, an area Mayor Osmeña has wanted to 'clean up' for years.[8]

It is possible to disagree with the effectiveness of mandatory blood testing in Cebu – as a small minority of AIDS activists in Cebu do – but it is also true that the new policy of the DOH was largely a policy response to the Manila situation where such initiatives were doing little to slow the spread of HIV. Furthermore, Mayor Lim's 'clean-up' campaign had forced the sex industry to move to other cities or go underground, thereby impeding the control, and thwarting the targeted efforts, of the DOH and NGOs working in the area. But Cebu officials saw the Manila policy as infringing on their authority, and nowhere did the tension between the national and local governments reveal itself as much as it did on the issue of mandatory testing and female sex workers. While mandatory testing was seen as an opportunity to assess the HIV/AIDS situation in the Cebu, the new policy was thought to undermine the authority of the City Health Department.

The City Government in Cebu is against prostitution, but Mayor Osmeña has claimed that 'the city health office is allowed to pretend that it is legal if only to pursue its mandate of delivering health services to all city residents'.[9] This conservative position on individual rights versus public health became important in the debate on mandatory testing. In 1993, revisions to the Sanitary Code proposed that the City Health Department be mandated to test all sex industry workers in addition to employees of beauty parlours (i.e. homosexuals) for HIV/AIDS (Figure 5.3).[10] Concerned about the ramifications of rejecting the national policy, the City Government consulted several lawyers on how to deal with the issue. These 'legal experts' suggested that women in the sex industry should be tested for the benefit of the larger community, and that mandatory testing was not necessarily an infringement on their human rights since they had waived some of their constitutional rights by working in the sex industry anyway.[11] Not only were these women perceived to have fewer constitutional rights, they were considered foolish to contest blood screening since they already submitted to 'degrading internal examinations',[12] and were described on many occasions as being willing participants in the procedure.[13] While these representations miss the point of the national policy – that mandatory testing is ineffective due to window periods, and discriminatory because it induces blame and stigma – they reveal how popular perceptions helped to blame these women for the spread

(a)
(b)

Figure 5.3 Mandatory HIV/AIDS testing in Cebu City. During 1993, there was much debate as to whether the mandatory blood screening of commercial sex workers and those employed in beauty parlours was an infringement on their human rights. Health Secretary Flavier spoke out against Cebu's proposed policy, but local officials declared their autonomy under the Local Government Code.

Sources: (a) *Sun Star Daily*, 6 December 1993.
(b) *The Freeman*, 9 October 1993.

of HIV. In the 1990s as in the 1930s, women engaging in sex work are perceived as the 'source' of disease, requiring control and/or isolation.

The City Health Department continues random and voluntary blood screening amongst Cebu's 'high risk groups', including sex workers and IV drug users. In April 1994, for example, the Health Department detected two HIV positive people who obtained the virus through IV drug use, and sought to use these cases as a mandate for further testing in the IV drug using community. But the language used to describe these interventions had unmistakably changed. As one news article noted: [14]

> The Department of Health will start here a campaign that will provide periodic HIV tests to two high-risk groups – the intravenous drug users and commercial sex workers . . . [The national coordinator for the DOH AIDS programme] said the periodic HIV testing among the high-risk groups here will be 'strictly voluntary'. All testings will be treated with the utmost confidentiality, he said. The DOH will coordinate with non-governmental organisations . . . to gain access to drug users and CSWs and engage them in counselling sessions for voluntary testing.

Although the language of high risk groups is maintained, the emphasis on voluntary testing and confidentiality may mark a new direction in the way officials in the City Health Department conceptualize testing. It is perhaps through these sorts of processes that the one persistent absence in this account of mandatory testing can be redressed: that is, the voices of women in Cebu's sex industry.

City politics and AIDS

As Brown (1995) has suggested, city politics mediate the local–global process of AIDS prevention activities, and in this case have played a role in the construction and reproduction of stereotypes of prostitution and HIV/AIDS. Nationally, the institutional and public emphasis on testing 'high risk groups' in certain geographic areas reinforced perceptions of sex workers as diseased, and placed the blame for the spread of the virus outside the country. While the policy to ban mandatory blood testing sought to abandon these caricatures, because the policy was largely a response to the situation in Manila, its acceptance in other parts of the country was resisted. This was certainly the case in Cebu, where a strategic emphasis on sex workers and foreigners, and most recently, on a newly designated group of IV drug users, was considered an appropriate response to the local situation. While this response reveals contemporary anxieties about the new global links in Cebu's economy, as well as old anxieties over social problems in Kamagayan, it also reproduced historical representations of women in Cebu's sex industry as diseased.

The major advocates of women in the sex industry have historically been feminist-nationalist organizations who view prostitution in relation to capitalist development and through a neoMarxist framework. Many of Cebu's NGOs continue to view prostitution from this position, primarily due to the unequal distribution of wealth in the city. While this perspective is useful in highlighting the uneven and neoimperial tendencies of global capital, it also tends to portray women in the sex industry as victims. When combined with the risk group mentality of public health institutions, these women are represented as victims of unequal structures of power and wealth *and* a deadly virus. Contemporary global imperatives have brought new perspectives to the debate, however, and analyses that emphasize the determined nature of the lives of prostitutes have been resisted by health-based organizations. These organizations emphasize empowerment in terms of sexual health, and tend to collaborate with the DOH to address issues such as discrimination. In interrogating how the 'risk group' label encouraged stigma and blame, for example, the Philippine National AIDS Council – composed of various representatives from government, non-government and transnational organizations – sought to redress this situation. The result was the Manila DOH policy which sought to ban mandatory testing.

While Cebu's officials may be correct in assessing the policy as Manila-focused and influenced by a global agenda, it is also true that the debate on mandatory testing in Cebu was also about who should be in a position to determine health policy. While recent decentralization efforts offer the opportunity for radical and plural democracy in this sense – an opportunity for local self-determination – a point must be made on where the local perspective was situated. While Cebu's officials were arguing against the national policy on the basis of their own autonomy, the issue of self-determination remained only at this level. Sex workers were not consulted on the issue; they were represented as either willing participants or as having waived their constitutional rights anyway. This was not the case in Manila, where genuine attempts were made to have sex workers represented on the National AIDS Council. Indeed, such representation reflects the imperatives of the international prostitutes rights movement, as well as the changing perspectives on how to address development issues in the Philippines. In the next chapter I examine the political tensions among and between NGOs, and how global perspectives on sex work have challenged historical representations of sex workers in the Philippines.

6

PROSTITUTE VICTIM/SEX WORKER AGENT

The global discourse of NGOs

Representations of prostitution in Southeast Asia have undergone transformations during the 1990s. The activities and discursive practices of development organizations – particularly foreign aid agencies and NGOs carrying out HIV/AIDS education projects – are in the process of unsettling the dominant representation of the prostitute as 'victim'. Historically, prostitution has been situated within the political economy of colonialism, militarism and sex tourism, and women working in the sex industry have been cast as victims of these structures. In the effort to slow the spread of HIV, however, and with the influence of international organizations, the types of intervention programmes for these women and the language used to describe prostitution are changing. Because 'empowering' women to protect themselves from HIV/AIDS is considered an urgent priority, perspectives that characterize women as victims have been challenged by an agency-centred, participatory approach to education. This perspective introduces a new prostitute subject: a 'sex worker' agent capable of managing her own destiny. For this reason 'sex worker' and 'prostitute' are used strategically throughout this chapter. Although 'sex worker' has a more specific meaning associated with debates on prostitution in the West, it is now legible in the discourse of government and non-government organizations in Southeast Asia to specify a particular kind of prostitute subject.

Many writers problematize the historical construction of prostitution in academic literature and government documents, arguing that it victimizes women, denies agency and fails to recognize the variability of experience.[1] In contemporary debates on prostitution and AIDS, there is conflict between those who privilege the free will of women to enter prostitution (agency) versus those who privilege the more deterministic constraints (structure) that make prostitution a job opportunity for women. This structure/agency binary is particularly evident in Southeast Asia where a forced/voluntary theoretical framework permeates discussions of prostitution. The majority of activists – both Asian and Western – argue that poor women have no choice but to work in the sex tourism industry, or to migrate overseas as prostitutes, where they can earn several times what they could earn in other professions. Although the economic infrastructure of the sex industry is crucial to the dynamics of sex tourism, activists articulating this view tend to stereotypically represent Asian women as 'passive and exploited victims' in a neocolonial slave trade.[2] Activists who do acknowledge the difference between forced prostitution and employment in tourist bars such as those described in Chapter 3 are regularly suspected of miscomprehending the complexities of prostitution or identifying too closely with the Western ideology of sex work. That is, of accepting sex work as a term used to describe one form of women's work, and to address basic human rights and working conditions in the sex industry.[3]

Doezema (1998: 42–7) has interrogated debates around forced and voluntary prostitution in Asia, arguing that the forced/voluntary dichotomy maps on too neatly with an innocent/guilty moralism and merely reproduces the whore/madonna division long associated with the category 'prostitute'. Moreover, she argues, an emphasis on forced prostitution has the tendency to ignore the rights of sex workers who work in the industry voluntarily (i.e. because they are 'guilty'). International networks such as the Coalition Against the Traffic in Women (CATW) consider all prostitution a form of violence against women, for example, as well as a violation of women's human rights. While their primary goal is to prevent women from being forced into prostitution – a goal few activists would contest – CATW's activities have the tendency to conflate all sex industry jobs with sexual slavery. Their research and advocacy work circulates internationally, and essentializes the experiences of women coerced into the sex trade to represent all forms of prostitution. These views are significant because they have the ability to influence international documents such as those generated at the 1995 Fourth World Conference on Women in Beijing. While these documents initially defined prostitution as a form of violence against women, organizations such as the Network of Sex Worker Projects (NSWP), an international sex workers' rights group with strong links to the Asia-Pacific region, lobbied the conference organizers to specify when they were discussing 'forced' prostitution (Doezema 1998). Although NSWP may have qualified discussions in Beijing, because there was an emphasis on what constituted a form of violence against women, the rights of sex workers who enter sex work voluntarily were ignored. Activism around sex work as a form of work as well as an issue of rights has occurred more frequently at international HIV/AIDS conferences than at conferences about women.

Assumptions about agency have implications for the types of programmes that empower women to protect themselves from HIV, particularly in terms of community participation in AIDS education projects. These assumptions are important because they have the ability to initiate discursive practices at global conferences, in organizations such as CATW or NSWP or in government HIV/AIDS prevention policies. Discursive regimes are not only ideological deliberations since they have concrete effects: they can sanction or contest mandatory blood testing policies as well as influence what kinds of sex worker 'projects' are funded by international development aid agencies. In this chapter I examine how these discursive regimes influence discussions about prostitution, sex work and AIDS in the Philippines. Rather than portray organizations articulating views of women in the sex industry as prostitute victims or sex worker agents as having the right or wrong story, however, I situate their perspectives within a social context that gives meaning to their own lives and professional situations.

Representing sex work in the Philippines

The dichotomy between victim and agent paradigms has been evident in debates on prostitution in the Philippines. There is a tension between feminist-nationalist NGOs that theorize prostitution as a problem of political economy, and community-based health NGOs that deal primarily with empowering women to protect themselves from HIV. This tension is partially due to the increased profile of health-based organizations in an era of HIV/AIDS – they tend to be favoured by Western development aid agencies and have actively collaborated with the government to develop national education strategy – but also due to their differing approaches to prostitution as a 'social problem'. While feminist-nationalist NGOs seek to reveal and transform the structures which facilitate the entry of a disproportionate number of women into the sex industry, health NGOs have been more concerned with the use of 'participatory' development techniques to improve the status of community health. Here I draw from the experience of some of these NGOs, highlighting how assumptions about the agency of women in the sex industry have changed since the early 1990s. While these debates could be contextualized by interventions from the international prostitute rights' movement, they must also be situated within broader discussions of the Philippine Left.

Before discussing these NGOs, however, I give a brief overview of the dominant representations of prostitution over the past two decades, accenting the important role of the American military bases, and the institutionalization of sex tourism and sex trafficking in debates on prostitution. This situates, and is followed by, the perspectives of three NGOs currently debating sex tourism in Cebu. These NGOs are variously involved in articulating views based on the political economy of prostitution, a perspective enmeshed with local political debates addressing the negative effects of rapid urban growth in the city. This perspective is contrasted to that of NGOs working with sex workers in Manila, Angeles and Davao. These organizations are reformulating their stance on the role of political economy, and offer an interesting translation of the 'sex worker agent' model put forward by various factions of the international prostitutes' rights movement. I conclude with thoughts about how the situated and shifting perspectives of NGOs might help dislocate the hegemonic 'victim' subject and, in turn, how this might form the foundation on which new identities for sex workers could be constituted.

In 1985, Azarcon de la Cruz (1985: 42) suggested that widespread prostitution in the Philippines stemmed from the following causes: an export-oriented economy which strengthens the dependence on foreign investments and loans; an unjust tenancy system displacing the peasantry; uneven development in the rural and urban areas; the presence of US military bases and its built-in R&R industry preventing people from developing other sources of income; foreign loans spent on tourism and infrastructure projects resulting in devaluation and galloping inflation; a crisis in leadership maintained by militarization and a growing dissident movement especially in the countryside; and a neocolonial and semi-feudal culture and orientation that peddles women as commodities and propagates their status as inferior to men, especially to foreign men.

These elements quite precisely situated prostitution within the political economy of colonialism, militarism and sex tourism, and dominated the discourse of prostitution in the media and among activists and academics during the 1970s and 1980s.[4] The American military bases, the international sex traffic and sex tourism were instrumental in popularizing these issues, and in shaping the way prostitution was constructed both within the Philippines and internationally. Although the American bases dominated the discourse on prostitution in the 1980s, their departure, combined with persistent demands of the World Bank and International Monetary Fund to integrate into the global capitalist economy, changed the focus of debates to sex tourism, the traffic in women, and most recently, aesthetic issues relating to the negative image of Filipinas abroad. These four themes will be discussed in turn.

First, the important debate of the 1980s, and until the departure of the American bases in 1991, revolved around the proliferation of bars catering to American servicemen in Angeles and Olongapo. Women working around the bases were seen to epitomize the country's problems, with urban and rural poverty forcing young girls into prostitution to support their families, where they had little choice but to succumb to the whims of American men. This representation was a means of popularizing the anti-colonial, anti-base sentiment rife through the Philippine middle class (Ryan 1991), and prostitution was seen as a metaphor for the broader relations between the Philippines and the United States (Miralao *et al.* 1990; Enloe 1989; Sturdevant and Stoltzfus 1992).

Second, many claim that sex tourism was institutionalized during the time of President Marcos when the government defined and promoted a 'hospitality business' to attract foreign tourists. In 1973, women's groups from Seoul to Tokyo staged demonstrations against sex tours throughout

Southeast Asia, and in 1981 Filipina women's organizations staged political protests highlighting the structural inequalities making sex tours possible for Japanese men (Mackie 1988). Throughout the 1980s and 1990s, and as a result of the vicious circle of indebtedness to the International Monetary Fund and World Bank, sex tourism has more recently been perceived as a product of a foreign-dominated economy that relies on tourism dollars to repay debts (de Leon *et al.* 1991; Azarcon de la Cruz 1985). In the case of Cebu, for example, prostitution is highlighted as a problem of government policies which attract foreigners to the city, symbolized by the construction of an international airport. In their quarterly feminist journal, the Laya Women's Collective (1993: 1) stated:

> The tale of prostitution is ever present. Today, however, prostitution has been stretched to its most exploitative scale and form with government's implementation of 'tourism' and 'rest and recreation' industries. A case in point is Cebu . . .

Third, while sex trafficking usually implies forced sexual slavery (De Stoop 1994), in the Philippines it is understood as facilitated by institutionalized labour migration (Fernandez 1993). As depicted in Figure 6.1, poor Filipinas are often perceived as grasping the double-edged sword of overseas employment, only to be maltreated or end up in prostitution. Almost 500,000 women migrate to other countries annually, primarily to work as domestic helpers, nurses and entertainers and, according to David (1991: 16) 'every other Filipino in Japan today is a young woman, between 15 and 24 years of

Figure 6.1 The double-edged sword of overseas employment.

Source: *Philippine Daily Inquirer*, 19 September 1993.

Figure 6.2 Negative images of Filipinas abroad.
Source: *Sun Star Daily*, 26 August 1993.

age, has violated the terms of her visa (either by working without permit or by overstaying), is therefore deportable, and makes a living as an entertainer'. This popular representation amongst Filipino activists – the *Japayuki* – is of a poor woman who travels to work in one of the bars in Tokyo and, once she realizes her employment is nonexistent or that her visa is invalid, is coerced into prostitution (Mackie 1988; Ballescas 1992). Annual remittances from overseas contract workers were estimated at US$4.3 billion in 1996 (May and Mathews 1997: iii), and this representation is used to highlight the negative consequences of government imperatives for foreign exchange, and government support for overseas workers is seen to legitimize prostitution.

Fourth and most recently, however, there has been concern about the negative stereotypes of Filipinas overseas. Because the 1993 scandal of Filipina models and actresses engaging in high class prostitution in Brunei – gratuitously dubbed the 'Brunei Beauties' case by the local press – refuted the point that all women are economically deprived and forced into prostitution, the focus of this debate was therefore on the negative images of Filipinas abroad (Figure 6.2). These stereotypes are said to encourage the mistreatment, rape and/or the sexual slavery of Filipinas in foreign countries, and the Brunei Beauties case prompted a congressional inquiry into precisely these issues (Jumilla 1993). While some criticized the inquiry and excessive media attention given to the case, others remarked that it had done a 'service' to Filipinas, changing the stereotype to 'expensive whore, sought after by royal studs' (Cacho Olivares 1993).[5]

As the Brunei Beauty case demonstrates, however, the conditions under which women enter prostitution are highly variable. The dilemma is that despite constraints to alternative development, the privileging of the structural aspects of prostitution minimizes the complexity of life choices and agency available to women in the Philippines. By simplifying prostitution to an uncomplicated relation of domination, women are cast as victims, a portrayal that inadequately reflects the everyday experience of sex work.

Cebu's sex industry: The discourse of NGOs

In the early 1990s, there were three major NGOs involved in constructing prostitution as a 'social problem': End Child Prostitution in Asian Tourism (ECPAT), the Women's Resource Center of Cebu (WRCC) and the Visayas Primary Health Care and Services (VPHCS). I briefly describe these organizations to examine their representations of Cebu's sex industry, and more specifically, how government policies that encourage rapid urban growth in Cebu have helped these NGOs to articulate a political economy perspective. The only organization with a 'project' for sex workers was VPHCS, but ECPAT and WRCC are described here to address the issue of NGO identity; that is, to highlight how NGO representations of prostitution are integral to the articulation of their own position within political debates.

End Child Prostitution in Asian Tourism (ECPAT)

ECPAT-International was founded in 1991 and has support offices in 22 countries. The organization's prime objective is to end child prostitution in Asia, and particularly prostitution catering to foreigners. ECPAT-Philippines is based in Manila, and does advocacy work for the country's estimated 60,000 prostituted children (under the age of 18). The organization's major tasks are: (1) to educate the public on the human and social consequences of child prostitution and how a sex-related tourism development programme has partly triggered its growth; (2) lobbying and advocacy work for the enactment of laws protecting the rights and welfare of children, and prosecuting paedophiles; (3) campaigning against tourism policies that perpetuate prostitution; (4) networking with and providing support services to organizations involved in helping prostituted children; and (5) documenting child sexual abuse cases to facilitate the delivery of aid to those at risk and the prosecution of abusers.

During 1992, ECPAT-Philippines began a research project on child prostitution in the Visayas (the central region of the Philippines), and Cebu was one of the three cities chosen for investigation. ECPAT's incentive came from reports in the local and national press regarding the upsurge of sex tours to the region, and after an initial visit to Cebu, ECPAT sponsored an in-depth study of the sex industry in Cebu City and province. The study was undertaken in 1993 and forms the basis for ECPAT's campaign against child prostitution in the Visayas (see ECPAT 1994). ECPAT's (1994: 11) study was a description of child prostitution in Cebu which highlighted that child prostitution 'should not . . . be treated apart from the economic, social, cultural and political environment prevailing in any given country'. Given this position, ECPAT devotes much attention to how the economic development strategies of the Marcos and Aquino regimes aimed at attracting foreign investors and tourists, and how President Ramos' vision of 'Philippines 2000' – that is, Ramos' vision of the Philippines as a newly industrialized country or NIC in the year 2000 – conforms to, and perpetuates, this strategy.

According to ECPAT (1994: 12), Ramos' Tourism Master Plan recommends the promotion of:

> (1) a niche and mass market approach based on a mix of special interest tourism and vacation resort market development oriented to both domestic and foreign markets; (2) a cluster development approach based on a gateway and satellite destination concept; and (3) a balance between domestic and foreign markets.

Not surprisingly, ECPAT relates these strategies to prostitution, where special interest tourism becomes 'inviting the *mass market* of [foreign] paedophiles', and a cluster approach to development is understood as extending prostitution to more areas throughout the country. These strategies are seen to encourage sex tourism throughout the Philippines, thereby increasing the spread of disease and related health care costs. Furthermore, such development is seen to encourage women to seek the 'easy money' of sex tourism in large cities, thus facilitating the migration of women from rural areas.

ECPAT's also addresses Japan's involvement in Cebu's economic and tourism development. Many firms in the adjacent Mactan Export Processing Zone (MEPZ) are Japanese-owned, for example, and these firms are represented as exploiting the female labour force – primarily through low wages.[6] In terms of tourism development, the study notes that Japanese-owned hotels and beach resorts work to the disadvantage of local communities, because such development converts agricultural land, displaces farmers and limits their source of income. Finally, ECPAT records the number of *karaoke* bars that exist in Cebu and cater to Japanese clientele, and relates this to a general enticement of foreign investments by the government. The links ECPAT makes between the MEPZ, Japanese tourism and agricultural land are persuasive, and reflect a commitment to contextualizing the current changes in Cebu's political economy within broader political debates.

Tourist development has a long history of being tied to political agenda and economic strategies and is often pursued with little concern for its socioeconomic and cultural effects (Richter 1989). Many individuals and organizations blame tourism, combined with the uneven distribution of wealth, for the magnitude of the sex industry in Cebu (de la Cerna 1992; WRCC 1993a). NGOs such as ECPAT articulate this view and implicate Philippines 2000 as perpetuating the social inequalities of the past, without addressing the real issues of uneven development. In the process, however, ECPAT offers only one cartography of meaning and power which is populated by a particular type of prostitute subject.

The links ECPAT makes between sex tourism and the spread of disease, and between 'easy money' and migration are partial at best. They conform to stereotypical images of prostitutes as diseased, are saturated with moralisms, and marginalize a discussion of the multifaceted reasons why women might migrate to cities. Recalling the story of Virgie (Chapter 4), migration was primarily about oppressive family responsibilities, and 'easy money' referred to the reality that her lack of education formed an impediment to gaining employment elsewhere. This is not to assert, however, that the stereotypical images ECPAT produces are erroneous, or that they necessarily limit the types of prostitute subjectivity. It is the rather different assertion that the identity of 'prostitute' ECPAT seeks to project is enmeshed in the national polemic of Philippine economic development.

Women's Resource Center of Cebu (WRCC)

The WRCC was established in 1991, and serves as a resource centre for women's organizations in Cebu. Its objectives are: (1) to advance the full participation of Cebuano women in the establishment of a society founded on equality, freedom, justice and peace; (2) to answer the specific needs of women from different sectors in cooperation with institutions, community and volunteer groups concerned with programmes on women's development; and (3) to render services to women from different sectors in cooperation with programmes on women's development. The prime components of the WRCC's programme include education and training, research, the production of a newsletter, maintaining a library and data bank, and advocacy work which includes both strengthening existing organizations and sponsoring various symposiums on gender-specific and other relevant issues.

The WRCC primarily coordinates with urban poor and peasant women's organizations. Although the WRCC does not have any projects directly relating to prostitution, because they are affiliated with GABRIELA – an umbrella organization of women's groups based in Manila – they usually brief visitors on prostitution in Cebu. Visitors include researchers such as myself, in addition to national and international organizations working with women generally, or prostitution in particular. Briefings usually include an outline of government development plans, visits to WRCC's corresponding rural and/or urban communities, and what they call an 'exposure tour'. An 'exposure' is a night visit to a selection of Cebu's sex industry establishments, where visitors who would otherwise not enter such establishments can see at first hand what goes on. The images and emotion mobilized by these visits are then connected to the WRCC's political issues.

During the early 1990s, WRCC became concerned with the government's Metro Cebu Development Plan (MCDP). Under the MCDP, the whole of Metro Cebu was declared a non-agricultural area despite the existence of approximately 40 mountain *barangays* (suburbs) where agriculture was practised. In these farming *barangays*, second- and first-class housing estates were either proposed or under construction in the upper (mountainous) regions, and these developments were displacing poor communities. Because the developers were also involved in tourist resort development on the neighbouring Mactan Island, WRCC's advocacy programmes linked this analysis to government programmes which were encouraging tourism development in Cebu. Institutionalized tourism development in Cebu did not come into full swing until the time of President Aquino (i.e. 1986), and has rapidly expanded to provincial areas since the late 1980s. Such development, according to members of WRCC, displaces rural communities – in much the same way as residential development – and with the MCDP, is met with the approval and encouragement of government. In the case of resort development on Mactan, not only are farmers displaced, but few jobs are provided for the displaced population because resort employees are usually college educated. Under these conditions, WRCC claims that women are in a situation where they are victims to tourist development. Many women end up in prostitution.

WRCC publishes a newsletter, *Bakud*, which features articles on how various development issues, including tourism development, affect women. In the November 1992 issue, for example, an article on the 'women situation' in Cebu province discusses what is referred to as 'women in the claws of tourism'. The article states: 'Part of selling Cebu to foreign investors and tourists is the prostitution and trafficking of women . . . Tourism has encouraged the proliferation of sing-along bars, hotels, entertainment establishments and beach resorts' (WRCC 1992: 10). As in the case of ECPAT, a strong link is made between tourism development and prostitution, and the blame is placed on government policies relating to tourism and foreign investment.

In another issue of *Bakud*, women in Cebu's sex industry are described as possessing a 'frankness that sounds vulgar to bourgeois ears' (WRCC 1993a: 6). The article portrays women in the sex industry as uneducated, poor, drug-users, victims of foreign sadists, sexual slaves and, most recently, susceptible to HIV/AIDS. A related article in the same issue begins: 'Myleene is a prostitute. And she hates it' (WRCC 1993b: 7). Although the articles present the life histories of women in Cebu's sex industry, these stories represent women as victims of a flesh industry created by outward-looking government policies; that is, their voice is heard in relation to, and substantiates, those interests. Yet at the same time, however, WRCC is utilizing these stories in order to articulate their own identity; that is, their position within local debates about control over local development, and particularly with reference to the MCDP.

Visayas Primary Health Care and Services

Since its inception in 1989, the Visayas Primary Heath Care and Services (VPHCS), a health-based organization, has assisted people's organizations in the formation and establishment of health committees in poor communities in Cebu province. An important part of its objective is to advocate the universal right to health, and highlight the structural problems of government which affect the health of urban poor communities. In the case of HIV/AIDS, an integral part of its information campaign is to highlight what VPHCS identifies three political issues potentially increasing its spread: (1) the presence of the American military bases; (2) the increased number of overseas contract workers due to the government's inability to provide jobs within the Philippines; and (3) the government's promotion of tourism, and therefore prostitution.

VPHCS began HIV/AIDS educational activities in 1991. As well as incorporating this aspect of health into existing programmes among Cebu's urban poor communities, some members also initiated seminars in the bars catering to Western men around Fuente Osmeña. This emphasis was derived from initial surveys which stated: 'Emphasis should be placed on how to protect oneself from the disease especially now that there is an increase in the number of sex industry workers as a result of their displacement from Olongapo and Angeles cities' (VPHCS n.d.). As discussed in Chapter 5, the influx of women from the American bases gave rise to increased public concern about Cebu's sex industry. Not only did the number of workers increase during this period, it was feared that these women were infected with HIV.

Despite VPHCS's usual objective of creating community-based programmes – where empowerment is derived through participatory processes – no attempt was made to facilitate the organization of the sex worker community.[7] Educational activities in the bars included basic information on HIV/AIDS, how it is transmitted, how to prevent it and a condom demonstration (Plates 6.1 and 6.2). As a member of the VPHCS stated:

> Because Visayas Primary Health Care is basically a community-based programme, we just chose two clubs to extend our programme. But we do not wish to concentrate on them because City Health is doing that already. I think they are doing a good job because whenever I go to these clubs to conduct lectures, the girls are quite knowledgable already.

Plate 6.1 HIV/AIDS education materials in a Fuente Osmeña bar. The information posted here describes what AIDS is, and the risks associated with different sexual practices.

Such seminars, primarily a one-way education campaign, were not dissimilar to the occasional educational programmes run by the City Health Department. Women employed in Fuente Osmeña establishments were able to demonstrate their familiarity with HIV/AIDS issues, and if they faced problems in terms of protecting their health and minimizing the risk of infection, the explanation was beyond rational/clinical educational information. Such seminars gave sex workers the status of objects and not subjects of the prevention of HIV/AIDS, and as Edna commented after a seminar we attended: 'AIDS is just a problem in my kind of job – I wonder if I should resign.' The difficulty in highlighting the political concerns of

Plate 6.2 Discussion of AIDS risks in a Fuente Osmeña bar. The VPHCS chose two bars in the Fuente Osmeña district to conduct HIV/AIDS education seminars. These seminars were primarily one-way information campaigns where women working in these establishments were given basic information about how to protect themselves from contracting HIV.

VPHCS (the American bases and sex tourism) within this context relates to the often opposing political views of organizations and their 'target groups'. Fuente Osmeña dancers, for example, do not conceptualize the American bases in negative terms.

The Visayas group accepts prostitution as a structural problem of government and sex tourism as a colonial legacy. The 'sex worker' becomes merely a victim of structures and a vector of infection, a 'target group' to educate. Such a representation poses a problem to empowerment in the broad sense, where AIDS is perceived as a 'problem' resulting from particular social structures instead of an issue around which sex workers can be mobilized to protect themselves from HIV. The cases of Kabalikat and Talikala, described below, reject this overly determined social context by using the agency-centred 'peer education' model, yet at the same time marginalize the issues highlighted by VPHCS.

The three NGOs described above have refrained from organizing women employed in Cebu's sex industry. Instead, prostitution is constructed in relation to local and national policies that affect Cebu's economic development. This lack of organization may relate to the more general problem of the increasing distance between the perspectives of Cebu's nationalist-inspired

NGOs and their corresponding urban or rural poor communities. Magno (1993: 11), for example, suggests problems arise due to the contrasting views over issues such as foreign investment and tourism development:

> [The] urban poor population is interested mainly in acquiring employment opportunities in the metropolitan economy rather than in agrarian reform. Politically, they have not been attracted to the nationalist/autarchic economic programs characteristic of the mainstream Left. The urban poor are disposed towards programs that welcome foreign investments since they promise immediate economic opportunities.

The current division in the Philippine Left is changing the nature of social movements. According to Magno (1993: 1), the dominant division appears to be between the 'reaffirmists' and the 'revisionists'. Magno's interpretation – although admittedly, more of an ongoing debate – has implications in Cebu, where rapid development is inspiring some NGOs, including those described above, to 'reaffirm' old doctrines. Reaffirmation requires more structural analysis, is based on resistance, and rejects collaboration with government because existing power relations have not changed. Under these conditions it is not necessary to hear the 'voice' of sex workers because, as implied above, to hear their voices may be contradictory to the political campaign.

During my first visit to Cebu, for example, I had the opportunity to spend time with an urban poor people's organization (PO) which had initially organized in response to the Cebu Business Park, a major redevelopment project in Uptown Cebu which threatened to displace their community. When one member of the PO discovered my research interests, she offered to take me to her community and introduce me to some of the women working in the sex industry, one of whom was her niece. When I arrived the following day, I found two women, each with a child on their breast, willing to answer my questions on prostitution. Both claimed to be 18 years of age, and to have worked in bars catering to Filipino men in the Junquera district. I took the opportunity to solicit vague life histories and find out more about their living conditions, but by the end of the interview realized that I had been coopted into sponsoring them to work in the sex industry again. We met the following week at the City Health Department, because I had agreed to pay for their health clearance certificates. We went through the bureaucratic process of obtaining chest x-rays, mayor's permits, photographs and internal examinations.

A few weeks later, I visited the bar they had applied to work in. One had commenced work that evening and was grateful for the help; she could now afford food for her child, clothing for her family, and was considering sending her brother to school. The other, as it happened, was under 18 years of age and had been denied a certificate. She had never worked in the sex industry before – despite the assurances of her aunt – and would be ineligible for another year. I returned to my field notes from the interview, re-reading her accounts of previous experience in a downtown bar in Cebu, and realized that the story had been concocted for the benefit of my research and her entry into the sex industry. I was disillusioned as to what reliable information was available, but in hindsight realize that the story told me much about attitudes towards prostitution not only within urban poor communities, but within POs as well.

Within the representations provided by Cebu's NGOs, women in the sex industry have been placed in literal and symbolic opposition to the forces of capitalism and neocolonialism, despite the possibility that the women themselves – or as suggested above, the urban poor community in general – do not oppose tourism development nor other strategies which encourage foreign investment. Through this representation, this strategic moment of closure, NGOs perpetuate the stereotype of prostitute identity as being borne out of capitalist social relations: it is therefore only a partial representation of sex worker identity. Yet NGOs often bound their categories of analysis while at the same time recognizing that, at the empirical level, such boundary drawing is inadequate. At the level they perform their own identities – whether this be at the scale of local, national or international discourse – these categories/representations may be entirely adequate for practical purposes, although they may not reflect the everyday lives of their 'target groups'. In the case of the NGOs described above, for example, their analyses are made in an effort to address the problem of asymmetrical power relations within Philippine society, and particularly as they connect to contemporary development issues in Cebu.

NGOs, HIV/AIDS and 'sex workers' in the Philippines

In 1987 there were seventy-nine recorded HIV and AIDS cases in the Philippines. The majority of these had been detected in women working as entertainers around the American military bases, and women's organizations involved in the anti-bases campaign – led by GABRIELA, an umbrella feminists organization – brought this to the attention of the national government (cited in Enloe 1989: 88):

> Who are the producers of AIDS in the Philippines? Why does prostitution exist and proliferate in the military bases and our tourist spots? The danger and damage of AIDS to women and the existence of prostitution are, in fact, crimes against women. We are the products, the commodities in the transaction . . . Who, then, we ask, are the real criminals of AIDS and prostitution? Indict them, not us.

From the mid-1980s to the early 1990s, HIV/AIDS was easily assimilated into the discourse of American bases, overseas contract workers and sex tourism. It was perceived as another product of a foreign-controlled economy, but one with the opportunity to directly affect all Filipinos. The construction of AIDS as a problem of foreigners thus inscribed the feminist and nationalist agenda, and persisted to do so until the early 1990s (Tan *et al.* 1989). By 1991 there were a total of 298 HIV and AIDS cases, however, and this combined with the nomination of a new and enthusiastic Secretary of Health inspired several AIDS education projects. These were carried out primarily by health NGOs rather than feminist or nationalist organizations. With these projects came the involvement of international HIV/AIDS education 'experts', and challenges to the dominant discourse in the Philippines.

Here I draw on the experience of Australian and Filipino NGOs that receive foreign funding for AIDS education activities. Initially, these NGOs had made contrasting assumptions about the agency of women working in the sex industry, with Australian NGOs asserting the agency of sex workers and Filipino NGOs asserting the more structural determinants of prostitution. While this contradicts the colonialist tendency where Western women author themselves as agents and Third World women as victims (Spivak 1987, 1988; Mohanty 1988), the dichotomous tendency itself has been a regular feature in descriptions of Third World women, encouraging alienation and distorting the complexity of experience.

The contrasting representations of Australian and Filipino NGOs relate to more generalized representations of prostitution in the West and in the Philippines. For example, while Western prostitutes have been represented as deviant, abnormal or as victims of a patriarchal society (Perkins 1991), such women were viewed as 'individually' pathological, and victimhood was personalized, individuated and not seen purely in structural terms. Emergent sex worker organizations also stressed the individual agency of prostitutes (Nelson 1985), while resisting the legal stigma and victim language which excluded prostitution from legitimate employment, and the medical discourse which allowed HIV/AIDS to reinscribe the 'pool of infection' label of earlier public health models (Jenness 1993). Most recently, prostitution has been linked to broader discussions of sexuality and sexual identity (Trumbach 1991; Weeks 1985, 1987), and this perspective is influencing not only the discursive regime of AIDS prevention, but the AIDS education projects carried out throughout Southeast Asia.

I begin this section with a vignette from Women's Education, Development, Productivity and Research Organisation (WEDPRO), because its internal discussions highlight the broader changes in the discourse on prostitution in the Philippines. These discussions highlight how individual members of NGOs are very aware of the constructed nature of their categories of analysis, and how these categories are at least partially linked to the desires of funding agencies. WEDPRO's experience is followed by that of Talikala and Kabalikat, who have been variously involved in the international HIV/AIDS discourse.

WEDPRO

WEDPRO has worked with women 'entertainers' around the former American military bases in Angeles and Olongapo for more than a decade. In 1990 it was commissioned by the Legislative-Executive Bases Council – a body constituted to address the issue of bases conversion after the servicemen left – to envision alternative employment, economic livelihood and development needs for women in the entertainment industry. The organization's views on prostitution were specified in its preliminary report (Santos in Miralao *et al.* 1990):

> The problem of prostitution is a complex sociocultural, political and economic issue which must be addressed in the overall context of genuine social transformation that takes into account the questions of national sovereignty and democracy and women's liberation from all forms of oppression and exploitation . . . if the bases are seen by a growing number of Filipinos as an encroachment of foreign powers upon national sovereignty, the question of institutionalised bases prostitution is a question of social justice and national dignity particularly of our women, our children, and collectively our people.

Following an extensive period of research in both Angeles and Olongapo, WEDPRO submitted a 'participative bases conversion' proposal to the Bases Council (see Lee and WEDPRO 1992). The proposal included educational and training projects for prostitutes, ensuring they would not be left out of the restructuring of the local economies.

At present, WEDPRO is continuing its work in Angeles, where sex tourism has undergone a modest revival, partially due to its proximity to Manila (where Ermita's bars have closed), but also because their proposal was largely ignored by government, and there was therefore a lack of education and training programmes for former entertainers. WEDPRO's major undertaking has been to assist the development of LAKAS, a sex worker cooperative involved in a livelihood project of food stalls. The stalls, which are strategically located adjacent to the bars, are seen as an opportunity for women to organize and discuss issues. Simultaneously, they provide alternative livelihood skills for women who choose to leave the industry. Increasingly, however, WEDPRO is looking to HIV/AIDS education activities. As Linda, a WEDPRO representative, said: 'Donors don't want to give money for livelihood projects any more, so we're thinking about an AIDS education project.'

Linda suggested that funding priorities for HIV/AIDS marginalize the role of socioeconomic programmes, which are now seen as moralistic and not helpful in preventing the spread of HIV. She recounted, as an example, a recent conference her colleague attended, at which a Western woman espoused the 'prostitution as free choice' perspective, claimed that these were her findings in both developed and developing countries, and argued that organizations such as WEDPRO, in proposing livelihood activities, implied that women should leave their work. This was frustrating and perplexing to the WEDPRO representative, who felt that accepting prostitution as a 'choice' peripheralized important feminist and nationalist issues. Given the complexity of the issue of choice itself, the debate merely became an articulation of opposed ideologies and sterile dichotomies. Nonetheless, a recent survey conducted by WEDPRO in the major urban centres of the Philippines incorporated these perspectives. Still grappling with the prostitution as 'free choice' perspective, Linda queried: 'But what do I do if the women tell me they enjoy their work?'

It is within this context that two Australian-funded organizations carrying out AIDS education activities for sex workers in the Philippines are described: Talikala in Davao and Kabalikat in Manila. Both organizations received funding under AusAID's 'AIDS Initiative' during the early 1990s, and were 'partnered' with the Australian Federation of AIDS Organisations (AFAO), an Australian NGO. AFAO had been contracted by AusAID to be partners with a number of Filipino NGOs working with sex workers and gay men (Altman 1991). Such partnering creates what AusAID perceives as a more democratic process of evaluation, while at the same time allowing for a productive exchange of ideas and experience. While it is true that Australian organizations have much experience with 'community-based' HIV/AIDS prevention programmes in Australia – particularly with sex worker and gay communities – they also have a culturally distinct view of commercial sex. It is this distinctly Western construction, enmeshed in Western legal and social theory, which comes into conflict with the Filipino perspective described above. Community-based programmes undergo a translation in this space, producing a new and hybrid 'sex worker' subject.

The issue of 'choice' in prostitution has played a role in discussions on prostitution in Australia, but the dominant discourse has historically focused on prostitutes as individually pathological, as well as victim to a patriarchal society (Perkins 1991). Throughout the 1980s, however, and in concert with an international movement which politically mobilized prostitutes, sex worker organizations sought to remove the 'victim' label through decriminalization and law reform. While similar cases are documented in the United States (Jenness 1993) and Europe (Pheterson 1989), Australian organizations have been particularly successful in working with state governments and obtaining funding for their efforts. Since 1985, for example, sex worker organizations in Australia have been successful in securing government funds for HIV/AIDS education projects.

AIDS education projects for sex workers in Australia have been inspired by activities in the gay community. One model in particular, 'peer education', has been adopted by Australian sex worker organizations. Peer education is a community-based model for AIDS prevention, encompassing recent advances in health intervention projects: the project must be designed, implemented and evaluated by the 'target' community, and through this participatory process, the community is empowered to safeguard their own health.[8] The centrality of community in AIDS education has been documented elsewhere (McKenzie 1991; Altman 1993) and will not be elaborated here. It is important to note, however, that the community-based peer education method worked well for some gay men, at least for the men who identified as gay and were part of an established network (magazines, newsletters, organizations) through which HIV/AIDS information could be channelled. Sex worker organizations that adopted this model emulated these networks, through organizing, producing newsletters and advocating legislative reform. The success of peer education in Australia inspired many to consider this model's effectiveness in Asia, given that Asia was the new 'epicentre' of the virus, and the site of the internationally famous sex tourism industry. Talikala and Kabalikat were two organizations that adopted the model.

Talikala

> Realising the miserable plight of hospitality women in Davao City, a Maryknoll lay missionary in 1986 took the initiative of forming a group composed of herself, a social worker, and a girl coming from the ranks of hospitality women precisely to respond to the deteriorating condition of women engaged in the flesh trade. At the very start the group was very much aware that their main task was not to tell these women to quit their jobs for such a task would require them to assume the state's inherent obligation to provide alternative jobs to women. On the other hand they understood that their mission was to assist women regain their self-respect and enable them to attain the power to dictate the course of their lives and shape their own destiny as human beings.
>
> (Desquitado 1992: 126)

This statement reflects an alternative type of programme being carried out by NGOs in the Philippines. It places the responsibility for prostitution squarely in the hands of the state, while at the same time acknowledging that asking women to leave a profession they entered for financial reasons would be counter-productive. While Talikala does not officially entice women out of the sex trade, it is insinuated that they *should* be in other jobs where they could find 'self-respect' and 'attain the power to dictate the course of their lives and shape their own destiny'. Nonetheless, the translation of the sex worker as agent is evident in this passage.

Since its inception, Talikala has aspired to empower women in Davao's sex trade through health issues. Its original programme of basic information dissemination was replaced in the early 1990s, however, when Cecile, a former Talikala member, participated in an AIDS 'study tour' in Australia. It was at this time that the organization took on a 'peer education' project. Cecile recalled the study tour as a turning point for the organization:

> They were really selling it to us, you know . . . Anyway, we were using an exhausting systems method, it was boring. We had to do something. There were the same girls, the same costumes, the same customers . . . Before we couldn't get the same participation and enthusiasm.

Talikala began the organization with a former sex worker as one of its three staff members, and has continued incorporating sex workers into the project. The peer group staff members – all former or present sex workers – have been responsible for collecting the data on the sex industry in the city (e.g. Talikala 1993). This core group has also been responsible for facilitating the formation of a self-help group, which is autonomous from the activities of the NGO, and headed by the former sex worker who began the NGO; that is, it is a branch of the organization strictly composed of sex workers. This group examines employment issues such as: working conditions, problems with customers and owners, and so on.

Talikala's primary goal is 'empowerment', and it is believed that this can be achieved by using the 'peer education' model. A current Talikala representative states that: 'Being a peer educator, the women themselves can relate freely with one another, because they come from the same work, they have similar aspirations and dreams. There are no barriers and no fears of being stigmatized.' The decision to adopt the peer education model, and the subsequent changes in the nature of programme activities, have encouraged Talikala to allow for a more 'participative' development project.[9] Sex workers are now involved in defining prostitution as a 'social problem', which is a progressive achievement in such a projects. Such an approach, however, denies the complexity and context of prostitution, in addition to the constraints to alternative employment such as a lack of skills, the social circumstances providing the personal context to sex work, and the lack of control over sexuality generally and hence the freedom to work or not work in the sex industry. Indeed, Talikala's activities highlight the difficulty in addressing the problems of political economy, while at the same time allowing a perspective from women in the sex industry.

Kabalikat

Kabalikat is a health-based organization that has been in operation since 1979 conducting community-based projects on oral rehydration, water and sanitation, family planning and more recently HIV/AIDS. Its first HIV/AIDS education activities began in 1989, and coincided with the identification of a number of women infected with HIV working in the sex industry. Because at that time only sex workers were being tested for HIV, it was not surprizing that Kabalikat decided to focus on these women. Its target group was the women in the tourist bars of Ermita, Manila. After a period of research in this area, and particularly after the executive director visited Australia on a study tour of AIDS education projects, the organization decided to use the 'peer education' model.

Kabalikat's methodology was to visit the bars and establish contact with what they described as the 'sex worker community'. Kabalikat's delineation of the community was what its members called the 'working system', the sex industry of Ermita, which included sex workers, managers and bar owners, clients, law enforcers and the relatives of the women. In practice, however, Kabalikat's primary focus was on sex workers. So sex workers were identified with what might not necessarily reflect their community – that is, their neighbours, family and friends. As Edna said:

> Even your best friend in the bar you cannot trust. Maybe one night she needs the extra (money) and she takes your boyfriend and tells him you've been going around with other men. Then they're going to the hotel and forget about you . . . Me, I just work there and come home to my house.

Kabalikat's framework, however, highlights the imperative of establishing community-based AIDS education projects. The use of the term 'community' is, however, problematic, since it presumes a common (sexual) identity. That community-based strategies work so well in homosexual communities is not surprising given that:

> [I]n reality the lesbian and gay community has always been a network of communities, primarily split between a political community which pursues an ideal of political power and representation, and the communities of the night, which encompass the genuine diversity which political activists seek to represent.
>
> (Alcorn in Watney 1989: 13)

While this may be the case for some men in the Philippines, it is less so for women working in the sex industry, who arguably have little common purpose and who lack an established network of organizations or newsletters.

In the rhetoric of development, however, community-based frameworks are privileged and there is little questioning of the existence or otherwise of a community from which base to work.

When the Mayor of Manila closed the bars of Ermita in his moralistic crusade, Kabalikat discontinued its original programme. Although a small handful of bars remained furtively in operation, the majority of women migrated back to their hometowns or sought employment in the new red light district only a few kilometres south. Those who remained were subject to harassment by the police in bars, or solicited business on the streets, and Kabalikat offered assistance to these women through monitoring police raids. Indeed, Kabalikat was frustrated by the moralistic crusade of Mayor Lim, which upset their delicately established peer education project. Kabalikat continued its commitment to sex workers by representing them on the government's AIDS Council, however, where they discussed issues such as mandatory testing. As one Kabalikat representative said:

> [Sex workers] are usually the ones infected more often than they spread it . . . these women do not choose to be in their profession . . . That is why we are hurrying up the resolution that makes no mandatory testing a national policy, so that all local ordinances can be revoked . . . Even if you force them to take the test: one, you cannot forcibly bring them to the facility; two, is the facility equipped to manage all the cases it finds positive whether for STDs or HIV? It does not put the responsibility on the person who puts his or herself at risk. It puts control over them by some other group.

These comments also highlight how HIV/AIDS is altering the discourse on prostitution in the Philippines. On the one hand, sex workers 'do not choose to be in their profession'. This view has a long history in the country, privileging the structural issues as described above. On the other hand, however, they should be 'responsible' for protecting themselves since they 'put' themselves at risk, without 'control over them by some other group'. One statement denies, and the other enhances, the agency of sex workers. Although seemingly contradictory, this represents a shift in the way NGOs conceptualize prostitution.

Importantly, Kabalikat's participation in local, national and international discourse is critical because they have been at the forefront of promoting 'peer education' within the Philippines. Kabalikat members participated in debates about Ermita, as well as discussions with the national government, and their experiences were often disseminated at regional and international HIV/AIDS conferences. It could be argued that the articulation of Kabalikat's identity as an NGO was a performance of the international discourse of HIV/AIDS prevention, yet this perspective would marginalize a discussion of the important local and national debates Kabalikat contributed to.[10] Indeed, Kabalikat's identity was fluidly constructed within the context of decentralization and division in the Filipino Left, where collaboration with government was a contentious issue.

Re-imagining the subject

The presumed universality of the concepts of prostitution or sexuality requires interrogation for them to be used in transcultural contexts (Caplan 1987) and particularly for HIV/AIDS intervention projects (Vance 1991; de Zalduondo 1991). Issues of translation and the differing interpretations of concepts such as 'peer education' are not easily separated from politics, and transcultural relations such as those evidenced in development aid – where foreign and local NGOs jointly create and evaluate projects – are entangled in a global structure which is hierarchical. This much is not at issue.

As the stories of Virgie, May and Edna illustrate (Chapter 4), women in the sex industry perceive a negotiated tension between their free will to enter prostitution and the more deterministic constraints that make prostitution a job opportunity for them. Organizations that neglect this complexity initiate discursive practices that deny the fluidity of lived experience, and women in the sex industry are less inclined to identify with their political projects. The tension between paradigms is most pronounced in the activities of NGOs involved in HIV/AIDS education, yet shifting the debate from no choice to free choice – and all the Western capitalist politics and ideology that go with it – would be a step in the wrong direction.

From a slightly different perspective, however, it could be argued that stereotypical representations deployed by NGOs are complex and ambivalent, and that it might be more productive to extend the critical and political objectives of representation to a new object of analysis (Bhabha 1994: 70). With this insight, it would be unproductive to describe Filipino NGOs as merely articulating discourse that stereotypes women as either victims or agents. It would be to deny a more complex account that emphasizes the negotiation involved in debates on prostitution and HIV/AIDS, and how these debates facilitate the articulation of particular types of NGO identity. It would deny, for example, that Cebu's NGOs are intervening in local political debates that address pressing issues of social inequality. It would also overlook how Kabalikat's appeal to the international discourse on HIV/AIDS has been instrumental in dislocating the 'victim' stereotype which dominated the discourse of prostitution for two decades, as well as their struggle to influence government policy. Most importantly, however, neatly categorizing NGOs as either articulating victim or agent paradigms denies the ambivalence of their identifications with women in the sex industry, as well as the possibility that the victim/agent debate itself could form the foundation on which new identities are constituted.

In his discussion of the polarization which often occurs in political struggles, Bhabha (1990) notes that the fixity of the difference between factions within a debate is often visible only at the surface. Indeed, allowing for the possibility of translation and hybridity within political rhetoric may be a means to acknowledge and overcome such binarisms. Bhabha's appeals to hybridity and the negotiation of difference are significant because 'in any particular political struggle, new sites are always being opened up, and if you keep referring those new sites to old principles, then you are not actually able to participate in them fully and productively and creatively' (Bhabha 1990: 216). NGOs, for example, have their own sites of power, meaning and identity from which they produce discourse. These sites are always shifting grids of meaning, and depend on whether their encounter is with women in the sex industry, the national government or with various international organizations. To describe these encounters as simple power relations – that is, to describe Filipino NGOs as homogenizing sex workers or foreign NGOs as deploying a colonizing discourse – would be to deny the *excess* which occurs in these encounters (Pile and Thrift 1995); that is, it would foreclose a 'third space' where it is possible to re-imagine the subject.

It is preferable to conceptualize HIV/AIDS, as well as its accompanying institutions, as providing the basis for the 'enunciative moment' to which Bhabha (1994: 36–9) refers. HIV/AIDS has unmistakably changed the way Filipinos conceptualize prostitution. NGOs in particular have incorporated this new dimension of the sex industry to their analyses of prostitution, and these discussions have played a role in defining broader political projects such as the scope of participatory development and the future of the Philippine Left. It is only through the intervention and emergence of this third space, this space beyond dualisms, that it is possible to re-imagine the prostitute or sex worker as different kind of subject.

7

CONCLUSION

> 'What is to be done?' must acknowledge the force of writing, its metaphoricity and its rhetorical discourse, as a productive matrix which defines the 'social' and makes it available as an objective of and for, action.
>
> Bhabha (1994: 23)

It is difficult to discuss sex tourism as part of a global economy characterized by unequal structures of power and wealth, while at the same time bringing to life the unfamiliar points of view expressed by women working in local sex industries. *Sex Work in Southeast Asia* acknowledges and transgresses this politics/everyday life dualism by highlighting the situated and strategic character of knowledge claims advanced by a range of people involved in the discourse of sex tourism. Rather than assess these claims on their ability to mirror the complexities of everyday life – a task requiring complex mediations – I have elaborated the 'third spaces' which allow particular accounts of sex work to be articulated. Indeed, it is only through acknowledging the contingent and dialogic character of representations of prostitution that the processes of displacement and realignment already occurring in debates about sex work and AIDS can be understood. This approach does not offer alternatives in terms of strategic political action, where self-conscious subjects resist set positions and identities of domination and subordination. Reinterpreting the encounter between the worlds of political discourse and everyday life does, however, offer the possibility of conceptualizing where a moment of political change might be located.

As highlighted throughout this book, women in the sex industry do not identify with representations that place them as victims of political economy, sex tourists or HIV/AIDS. Indeed, they perceive a negotiated tension between their free will to enter prostitution and the constraints that make this particular type of employment an opportunity for them. Moreover, sex workers do not conceive their encounters with foreign men in strictly oppressive terms. Dichotomous models of power – whether based on the colonizer/colonized or oppressor/subordinate binaries of nationalist and feminist discourse – are necessarily more complex, situated and strategic. Furthermore, the subjects of these models are capable of power and resistance in complex and unanticipated ways. In go-go bars throughout Southeast Asia the differences between white, Western men and Southeast

Asian women are worked out, but this negotiation is not based on a clear relation of domination. Men and women are more than oppressive sex tourists or passive Asian women, and the performance of stereotypical identities is primarily influenced by attempts to overcome uncomfortable processes of social and cultural identification – processes which are laden with, and conditioned by, contemporary moral and secular taboos.

This approach to understanding the relationship between sex workers and their customers has much affinity with postcolonial perspectives which seek to disrupt the nationalist ideologies that set up a binary relation between Third World/First World (Bhabha 1994). These perspectives depart from recurrent theories of underdevelopment or dependency and, according to Bhabha (1994: 173), recognize the 'complex cultural and political boundaries that exist on the cusp of these often opposed political spheres'. While the relationship between sex workers and their customers is the most detailed example given here, a similar argument is made about the relationship between Australian NGOs and their Filipino NGO 'partners'. To define peer education as a purely colonizing discourse, for example, would be at the expense of understanding the Filipino translation and use of the concept.

As this study shows, organizations with specific political agendas utilize dichotomous notions of power which are incommensurable with the lives of women employed in the sex industry. Due to the global concern with HIV/AIDS, these organizations include an increasingly complex web of international agencies (i.e. the United Nations, aid organizations from the West, etc.), national and local governments, and NGOs carrying out community-based HIV/AIDS education projects. Furthermore, the advent of HIV/AIDS has brought new voices to these debates, including those from the Western-based prostitutes' rights movement. Through examining the perspectives of local, national and foreign organizations – as well as the uneasy processes of social and cultural identification between them – it becomes clear that representations of prostitute victims/sex worker agents are also articulations of organizational identity. Because the images the produced by government and non-government organizations are as much about their own positions within debates about sex work and AIDS, it could be argued these images are politically strategic stereotypes – and are therefore fragile and ambivalent. While they might deny the lived experience of sex work, they can also be considered the foundation upon which new identities might be constituted.

Examining the ambivalence of these stereotypes, while at the same time linking AIDS prevention discourse to the contexts in which cross-class, cross-gender and cross-cultural identifications are constituted, is one way to reconceptualize debates on prostitution, sex tourism and HIV/AIDS. It becomes possible to examine how the sites of sex tourism (i.e. bars, red light districts, cities, entire countries) are differently imagined by women in the sex industry, Western sex tourists and organizations involved in HIV/AIDS prevention, while at the same time considering how the ideological divisions and differences between these groups might be revalued. Yet a pressing question remains: what is to be done? Does this approach acknowledge that the discursive categories articulated by NGOs and other organizations sanctioned by Southeast Asian and Western governments continue to control knowledge and meaning through the institutions and social relations concerned with development aid? Does it acknowledge that these categories help to constitute the lived experience of sex work, often in negative and stigmatizing ways? On the other hand, does a perspective that privileges multiple voices, ambivalence and ambiguity empower people to protect themselves from HIV? As Escobar (1995) and others have argued, deconstruction and reconstruction are projects that must be carried out simultaneously.

Theory, practice and postmodern development theory

> We cannot understand what is being proposed as 'new times' within postmodernism – politics at the site of cultural enunciation, cultural signs spoken at the margins of social identity and antagonism – if we do not briefly explore the paradoxes of the language metaphor . . . the language metaphor opens up a space where a theoretical disclosure is used to move beyond theory. A form of cultural experience and identity is envisaged in a theoretical description that does not set up a theory-practice polarity, nor does theory become 'prior' to the contingency of social experience. This 'beyond theory' itself is a liminal form of signification that creates a space for the contingent, indeterminate articulation of social 'experience' that is particularly important for envisaging emergent cultural identities.
>
> Bhabha (1994: 179)

Homi Bhabha frequently and critically analyses the dualism of theory and practice. He rejects the claim that current social theory is bourgeois, suggesting a space between theory and practice that examines their articulation in a productive relation of supplementarity (Bhabha 1994: 179). This approach has much affinity with the concept of 'praxis' as used in development theory over the past two decades, where theoretical discourse and contemporary political movements are seen to work semiotically. In writing about sex work and HIV/AIDS in Southeast Asia, for example, I have drawn on contemporary political debates within the instrumental world of AIDS prevention organizations, as well as the concepts and heuristics of new theories of subjectivity. While this provides a space for a diverse array of voices within a single text (e.g. sex workers, NGOs, social theorists, and so on), the production of my narrative, and the identities represented herein, are the product of an artificial closure. The identities of the subjects presented are contingent upon both instrumental and theoretical interests, as well as my interpretation of our realm of intersubjectivity. At the same time, however, and through acknowledging the 'force of writing', I would also maintain that this closure is contingent, open to interrogation and a way 'beyond' these very categories of identity.

The question 'what is to be done?' has always been an important question for my work. As discussed in Chapter 4, my empirical focus shifted from identifying with government and non-government organizations to sex workers/dancers during the course of protracted fieldwork. This was largely a response to what appeared to be the incommensurability of discourses espoused by these organizations, and the extreme difficulty in hearing the voices of sex workers outside such instrumental interests. Upon entering the world of writing, however, it became less important to enumerate the voices of NGOs and sex workers than to construct a narrative strategy where the emergence and negotiation of different and incompatible voices rendered the experience of ambivalence – including my own – transparent. This approach, inspired by Bhabha's (1994: 183) conception of 'writing aloud', allows the intentionality of my text to be negotiated retrospectively; that is, it allows the meaning and agency of the subject to be negotiated through a 'time lag' between my fieldwork and the production of discourse.

Through the production of this narrative, I demonstrate the contingency of debates about sex work and HIV/AIDS. In particular, I illustrate how the discourse espoused by NGOs is contingent upon a number of factors including their positions in local and national political debates, division within leftist movements, links to international discourse through funding agencies, and so on. Furthermore, sex workers' perspectives of their employment are also contingent, and depend upon their relationships in the bar, their home and family situations, the availability of other options and various other factors. More importantly, however, the identities of sex workers and NGOs are contingent upon each other in that the discourse and categories of NGOs – despite their ambivalence – help to constitute the lived experience of sex work. Contingency in this sense is not necessarily about indeterminancy, since there are always negotiations of power involved in representations of sex work. It is this contingent space, which holds the possibility of 'envisaging emergent identities', to which I will now turn.

A space for a speaking sex worker subject

The social construction of categories helps to constitute social identities, whether these identities are 'woman', 'dancer', 'sex worker', 'risk group' or 'victim'. It is important to understand how these categories intersect everyday lives, and how people resist, subvert or transgress them. In the case of sex workers, sexual and cultural taboos help to discursively produce these identities in ways that mark their bodies with sexuality, class, race and gender. The advent of HIV/AIDS has also re-marked sex worker bodies as diseased. If the performed identity of women in the sex industry can be understood as an effect of such categories, as Judith Butler (1990, 1993) argues, there is a need to move beyond portrayals which construct these women as victims. Indeed, agency is found in the articulation of identities, and not through the construction of a singular category or image, and imagining new grounds for a speaking sex worker subject also has implications for AIDS prevention activities. This perspective, I would argue, moves beyond the incommensurable discourse of AIDS prevention organizations. It is through acknowledging the contingency of categories to the performance of identity which helps to answer 'what is to be done?' types of questions.

The experience of social marginality holds the potential to transform the critical strategies that address sex work as a 'social problem'. Throughout Southeast Asia, however, there has been little space for self-authored sex worker identity. This is not to condemn those who speak on their behalf, or to propose that the concerns of sex workers are the only relevant problems. It is the rather different objective of including them in part of the dialogue. This approach is in concert with current writing in development theory (see, for example, Slater 1992; Marchand and Parpart 1995; Escobar 1995; Crush 1995). These authors draw on the insights of representational and poststructural theory, while simultaneously linking such insights to critical thinking on development issues. While these texts have been important in conceptualizing development as a 'discourse' generating its own institutions and social relations, they have also been important in stressing the significance of subjectivity and personal identity and how the voice of the 'other' might be integrated into development practice. More specifically, and as Marchand and Parpart (1995) argue, the voices of 'the oppressed' should be solicited in participatory environments and situated in specific contexts. In this way, the category itself – which implies victimhood – can also be addressed.

It is useful to return to the 'peer education' model within this context. While this type of participatory education offers the potential to articulate new identities, to date, there has been little room for sex workers to participate in envisioning new kinds of projects. Their participation is often limited to sharing HIV prevention information with co-workers, rather than managing the broader agendas of NGO activities. Some studies of peer education in Asia suggest that these projects reinforce the singular identity of these women as prostitutes as well as the stigma of sex workers as diseased and dangerous. Furthermore, organizations carrying out such projects can be more accountable to their funding agencies than sex workers (Murray 1996). It is true that the instrumental world of NGOs is influenced by the global imperatives of funding agencies, yet all is not doom and gloom.

The global discourse of HIV/AIDS and the introduction of peer education in Southeast Asia have successfully challenged the stereotype of the prostitute as (always) a victim. This in itself represents a significant shift in official attitudes regarding prostitution, and offers the foundation on which new sex worker subjectivities might emerge. Because the identity of women in the sex industry is enmeshed with the categories of AIDS prevention and feminist organizations, and because these categories themselves are in the process of transformation, it could be argued that peer education has already produced a political and plural sex worker subject. While the essentialized identity of a prostitute victim may have been called into question, another question must also be asked: does this subject speak?

Parpart (1995) has noted that: '[t]he possibility that Third World women know how to act in their own interest has largely been ignored'. For women in the sex industry, there has been a long history of socially inappropriate development interventions (e.g. rehabilitation, skills re-training) which have done little to empower or organize these women. In the case of HIV/AIDS education, there has also been a tendency to assume these women require technical information, although 'peer education' has gone beyond concepts of development which emphasize improving material conditions or technical assistance. Indeed, peer education has acknowledged that (a) women are sources of information; and (b) that empowerment should be based on processes that allow women to make sense of their own lives.

Poor women can contribute to feminist theorizing, and Marchand (1995) suggests that life histories are one way to situate and contextualize current development issues. Life histories should be considered more than background information, Marchand (ibid.: 68) argues, and can serve to contextualize a range of gender and development issues. While the limitations of using life histories is discussed in Chapter 4, life histories can be of importance to participatory development generally, and AIDS prevention in particular. Through this approach, women in the sex industry and NGOs may be able to find areas of overlap between their interests. Although HIV/AIDS currently overshadows other issues in the sex industry, and this is largely a product of global imperatives to slow the spread of the virus, this is clearly not the only issue. If sex workers are concerned about financial support for their Eurasian children, oppressive family responsibilities in rural areas or the quality of their working conditions, these issues provide an opportunity to address broader social issues/power relations in a way which directly relates to women in the sex industry. Furthermore, in places such as the Philippines, where Catholicism mediates the subject positions of women through dualisms such as good/bad and madonna/whore, NGOs need to define new discourses to address such issues of stigmatization for women to identify with their agendas. Indeed, NGOs need to find a form of practice that goes beyond the current rhetoric of participatory development.

Given that current debates on prostitution, HIV/AIDS and sex tourism are contingent, this also suggests they are open to change. Creating a space for a speaking sex worker subject – where the problems and difficulties of working in the sex industry are articulated by women in the industry themselves – is one way to question dualistic thinking in current debates. It is only when these women are able to address and advocate changes on their own behalf that counter-hegemonic discourse, and new identities, can emerge.

NOTES

INTRODUCTION

1 Virgie's life history is documented in Chapter 4.
2 I am referring to the maps in Odzer's (1994) publication *Patpong Sisters*.
3 I borrow this terminology from Jackson (1992).

1 RETHINKING THE PROSTITUTE SUBJECT

1 See Lloyd (1993). This work, originally published in 1984, discusses the Cartesian dualism of the Enlightenment in relation to earlier dichotomies in Western classical and medieval philosophies, as well as the gendered character of the mind/body split.
2 For a review of these concerns, see Bordo (1990).
3 Much of this work has been inspired by French feminists such as Kristeva, Irigaray and Le Doeuff. For a review of this work, see Grosz (1989).
4 Grosz (1994: 15) refers to egalitarian feminists as 'figures as diverse as Simone de Beauvoir, Shulamith Firestone, Mary Wollstonecraft, and other liberal, conservative, and humanist feminists, even ecofeminists'.
5 This group contains most feminist theorists, Grosz (1994: 16) argues, including 'Juliet Mitchell, Julia Kristeva, Michele Barrett, Nancy Chodorow, Marxist feminists, psychoanalytic feminists, and all those committed to a notion of the social construction of subjectivity.'
6 Here Grosz (1994: 17) includes 'Luce Irigaray, Hélène Cixous, Gayatri Spivak, Jane Gallop, Moira Gatens, Vicki Kirby, Judith Butler, Naomi Schor, [and] Monique Wittig.'
7 For a detailed discussion of COYOTE, a pioneering prostitutes' rights group, see Jenness (1993).

2 CARTOGRAPHIES OF DESIRE

1 See, for example, Treichler (1989); Watney (1989); Patton (1990); Crimp (1992); and Grover (1992).
2 Prior to the 1990s the sex industry was referred to as the 'hospitality industry' and was part of President Marcos' attempts to sanitize commercial sex in his tourism development plans.
3 I am making these statements as a counter-point to Foucauldian-inspired texts such as Stoler's *Race and the Education of Desire* (1995) and Said's *Orientalism* (1991), which have the tendency to portray the Orient as passive and female. I take this lead from Bhabha (1994) and his psychoanalytic renderings of the

colonial encounter. For critiques of Said, see Jolly (1994) and McClintock (1995).

4 For a discussion of how the peer education model has been deployed in Southeast Asia, see Murray (1996). A fuller treatment of these debates is provided in Chapters 5 and 6.

5 See Pile and Keith (1997) for discussions of geographies of resistance.

3 NEGOTIATING THE BAR

1 Cebu City is the second largest city in the Philippines with a metropolitan population of approximately 1 million people. The events described here are not fictional; they occurred in many of the different bars in which my research was carried out (between November 1992 and March 1994). The names of all people and places have been replaced with pseudonyms.

2 Cf. Bhabha (1994: 38, 66–7).

3 Bhabha's (1990) 'bar' is derived from the space between signifier/signified, and he is primarily concerned with 'culture' and issues of meaning in translation. I have borrowed this quote and highlighted the *bar* to spatialize this concept within the entertainment establishments of Cebu City.

4 Many authors, particularly within feminist and cultural studies, have sought to theorize this space. See, for example, Irigaray (1985); hooks (1991); Grosz (1994); Bhabha (1990, 1994); Keith and Pile (1993).

5 The work I refer to here is the well-known literature on the English working class (Willis 1977) and peasants in Southeast Asia (Scott 1985). Ong's (1987, 1991) important work on factory women in Malaysia also focuses on class-based resistance, but she brings issues such as indigenous religious systems and Japanese management techniques to bear on the concept of resistance.

6 See also Murray (1995).

7 'Social worker' was the term Edna used to describe a range of government and non-government professionals, whether involved in health, employment or community development issues.

8 For a review of some of these perspectives, see Enloe (1989), Miralao *et al.* (1990) and Sturdevant and Stoltzfus (1992).

9 Geographers have also discussed the phenomenon of personal comfort in terms of the production of private and public space (e.g. Valentine 1989), and gay and straight space (e.g. Bell *et al.* 1994).

10 A barfine is the amount a customer pays to take a woman outside the bar, although dancers often refer to their paying customers as barfines as well. Barfines in Cebu were approximately P300 during 1992–94, and were usually split 50/50 between the woman and the bar. A night's salary for dancing and entertaining customers ranged from P100 to P150. During this time the exchange rate was approximately US$1 = P25.

4 BEYOND THE BAR

1 Gregory (1994) uses the term 'cartographic anxiety' to describe the contemporary problem of representation in geography. Post-structuralist analysis of objective representations of space and place has inspired much discussion in geography (Rose 1993b; Duncan and Ley 1993), and I use 'ethnographic anxiety' to highlight these issues, as well as draw links between discussions in geography and anthropology.

2 Kondo (1990) provides an interesting discussion of experience as theory. See also Pratt (1993).

3 These stories were collected at intervals over the period from November 1992 to March 1994.
4 Cf. Haraway's (1988) parallel argument about 'situated knowledges'.
5 In Chapters 5 and 6 I anticipate and address absences perceived by government and NGO discourses.
6 My interviews with Virgie were never recorded because she did not like the concept of a recorded meeting. We always worked with pen and paper, and spoke a mixture of English and Cebuano (although, mostly English). Dancers often spoke to me in English which frustrated my attempts to speak Cebuano, but often said they wanted to practise their English and asked me to correct their grammar.
7 During 1993 this rate was approximately US$1 = 25 peso.
8 See Robinson (1996) for an interesting perspective on how Asia is entering the Australian imaginary through debates about Filipina–Australian marriages. See Wall (1983) and Tapales (1990) for discussions of Filipina mail-order brides.
9 *Di ba* is an expression that transforms a statement into a question.
10 In the Philippines older women are always addressed Manang, a title used out of respect for elders.
11 See Moselina (1979), Miralao *et al.* (1990), Buklod (1990), Ryan (1991), Lee and WEDPRO (1992), Stoltzfus and Sturdevant (1990), Sturdevant and Stoltzfus (1992) and Enloe (1983, 1989).

5 SEX WORK, HIV/AIDS AND BLAME

1 The United States and Australian Agencies for International Development (USAID and AusAID respectively) fund HIV/AIDS projects in both public and private sectors. In 1993, USAID and AusAID contributed 24 and 36 per cent of the DOH's budget respectively (AIDS Research Group, RITM 1993). In addition, AusAID has funded a total of A$1,664,616 for NGO projects from 1991 to 1993 (AusAID 1993).
2 'Manila health officer warns on HIV cases', *Sun Star Daily*, 23 October 1993.
3 See the 15 February 1994 special feature in the *Sun Star Daily* entitled 'Update on HIV, AIDS: Special report on the health problem'. Emphasis added.
4 See the articles 'Sex workers on the rise in Cebu' and 'RP flesh trade: Sex tours enjoy Cebu revival' in the 16 February and 21 March 1992 editions of the *Philippine Daily Inquirer*, respectively, and the article 'Caution: More CSWs here to lure men' in the 11 March 1992 edition of the *Sun Star Daily*.
5 '17 HIV victims found in clinic: All foreigners with 4 of them full blown AIDS cases', *Sun Star Daily*, 23 March 1993.
6 A summary of these events is provided in Mangubat (1993).
7 'Access to clean needles, syringes may cut AIDS risk, but won't it worsen drug abuse?', *Sun Star Daily*, 20 March 1993.
8 See the 14 July 1993 article in the *Sun Star Daily* entitled 'Crackdown on sleazy pubs seen as "clean up" vowed: Drive readied after SC upholds Manila ordinance'. When the Supreme Court upheld the Manila ordinance that banned all establishments being used as fronts for prostitution, Mayor Osmeña vowed to 'clean up' Cebu. This would entail more regulation for entertainment establishments, and a total liquor ban in Kamagayan.
9 'Approach to prostitution two-faced, says Tom', *Sun Star Daily*, 12 February 1993.
10 A large majority of employees of beauty parlours in the Philippines are homosexuals. For information of the mandatory testing debate for homosexuals,

see the 5 October 1993 article titled 'Proposal to test all gays for AIDS an insult to them' in the *Sun Star Daily*, and the 9 October 1993 Editorial in the *Freeman*.

11 'Mandatory HIV test on nightspot workers "not unconstitutional"', *Sun Star Daily*, 3 December 1993.

12 *Ibid*.

13 See the 18 and 19 November 1993 articles titled 'HIV test is part of social hygiene, says health official' and 'No ordinance requiring HIV test for bar workers', respectively in the *Sun Star Daily*.

14 'Gov't starts periodic HIV tests in Cebu', *Philippine Daily Inquirer*, 8 April 1994.

6 PROSTITUTE VICTIM/SEX WORKER AGENT

1 See White (1990); Truong (1990); Perkins (1991); Sullivan (1994); Lyttleton (1994); Kempadoo and Doezema (1998).

2 Doezema (1998: 42).

3 Kempadoo and Doezema (1998: 3).

4 See Moselina (1979), Lewis (1980), Doyo (1983), Aquino Sarmiento (1988), Tapales (1988), Enloe (1983, 1989), Buklod (1990) and Stoltzfus and Sturdevant (1990).

5 The Brunei Beauties case is only one example of concern for negative images and stereotypes of Filipinas overseas. Indeed, there is a growing literature on the construction of Filipinas overseas, particularly as entertainers in Japan (Tyner 1996) and as domestic workers in Hong Kong (Constable 1997).

6 For a review of the gendered impact of the MEPZ on women's employment, see Chant and McIlwaine (1995).

7 Whether women working in the sex industry can be described as a 'community' requires more attention. Some NGOs have adopted a community-based approach to HIV/AIDS education for sex workers in Manila and Davao, as discussed below.

8 Participatory development practice has been an ongoing discussion in the Philippines as well as in other 'developing' countries. The lack of progress in poverty alleviation during the 1970s gave rise to a more people-centred, basic needs approach during the 1980s and 1990s (for a review see Cohen and Uphoff 1980). Participation is a popular strategy employed by NGOs in the Philippines, and is applied in both rural and urban settings (see Osteria and Okamura 1986).

9 For a discussion of community-based, participative AIDS education projects, see Gorman and Mallon (1989) and Seeley *et al.* (1992).

10 For a discussion of issues relating to autonomy and co-optation, and how Filipino NGOs perceive their cooperation with government, see Constantino-David (1992).

BIBLIOGRAPHY

Abu-Lughod, L. (1990) 'The romance of resistance: tracing transformations of power through Bedouin women', *American Ethnologist* 17(1): 41–55.

Aggleton, P., Weeks, J. and Taylor-Laybourne, A. (1993) 'Voluntary sector responses to HIV and AIDS: a framework for analysis', in P. Aggleton, P. Davies and G. Hart (eds) *AIDS: Facing the Second Decade*, London: The Falmer Press, pp. 131–40.

AIDS Research Group, RITM (1993) 'HIV Infection/AIDS in the Philippines', Paper presented at a United Nations Development Program/Asian Development Bank workshop titled 'Study on the economic implications of the HIV/AIDS epidemic in selected DMCs', 8–10 September 1993, Manila.

Allison, A. (1994) *Nightwork: Sexuality, Pleasure, and Corporate Masculinity in a Tokyo Hostess Club*, Chicago: University of Chicago Press.

Altman, D. (1991) 'The primacy of politics: organising around AIDS', *AIDS* 5 (suppl. 2): 231–8.

—— (1993) 'Expertise, legitimacy and the centrality of community', in P. Aggleton, P. Davies and G. Hart (eds) *AIDS: Facing the Second Decade*, London: The Falmer Press, pp. 1–12.

Amante, I.D. (1993) 'Fear of side effects, not religion, prompts women to pick natural method', *Sun Star Daily*, 4 August.

Aquino Sarmiento, M. (1988) 'Love for sale (like cigarettes)', *Sunday Inquirer Magazine*, pp. 12–13.

AusAID (Australian Agency for International Development) (1993) 'Summary of on-going/newly approved health projects', unpublished report. Manila: Australian Embassy.

Azarcon- de la Cruz, P. (1985) *Filipinas for Sale: An Alternative Philippine Report on Women and Tourism*, Manila: Aklat Filipino.

Ballescas, R.P. (1992) *Filipino Entertainers in Japan: An Introduction*, Quezon City: The Foundation for Nationalist Studies Inc..

Bell, D. and Valentine, G. (1995) *Mapping Desire: Geographies of Sexualities*, London and New York: Routledge.

Bell, D., Binnie, J., Cream, J. and Valentine, G. (1994) 'All hyped up and no place to go', *Gender, Place and Culture* 1(1): 31–47.

Bell, L. (ed.) (1987) *Good girls/bad girls: Sex Trade Workers and Feminists Face to Face*, Toronto: The Women's Press.

Bell, S. (1994) *Reading, Writing and Rewriting the Prostitute Body*, Indianapolis: Indiana University Press.

Bhabha, H. (1990) 'The third space: interview with Homi Bhabha', in J. Rutherford (ed.) *Identity: Community, Culture, Difference*, London: Lawrence and Wishart.

—— (1994) *The Location of Culture*, London and New York: Routledge.

Bordo, S. (1989) 'The body and the reproduction of femininity: a feminist appropriation of Foucault', in A. Jagger and S. Bordo (eds) *Gender/Body/Knowledge: Feminist Constructions of Being and Knowing*, London: Rutgers University Press, pp. 13–33.

—— (1990) 'Feminism, postmodernism, and gender-scepticism', in L.J. Nicholson (ed.) *Feminism/Postmodernism*, New York and London: Routledge, pp. 133–56.

—— (1992) 'Postmodern subjects, postmodern bodies', *Feminist Studies* 18(1): 159–75.

Brown, M. (1995) 'Sex, scale and the "new urban politics": HIV-prevention strategies from Yaletown, Vancouver', in D. Bell and G. Valentine (eds) *Mapping Desire*, London and New York: Routledge, pp. 245–63.

Buckley, S. (1997) 'The foreign devil returns: packaging sexual practice and risk in contemporary Japan', in L. Manderson and M. Jolly (eds) *Sites of Desire/Economies of Pleasure: Sexualities in Asia and the Pacific*, Chicago: University of Chicago Press, pp. 262–91.

Buklod (1990) *Hospitality: What Price? The US Navy at Subic Bay*, Olongapo: Buklod Centre.

Butler, J. (1990) *Gender Trouble: Feminism and the Subversion of Identity*, New York and London: Routledge.

—— (1993) *Bodies that Matter: On the Discursive Limits of Sex*, New York and London: Routledge.

Cacho Olivares, N. (1993) 'Maceda, Coseteng did a "service" to Filipinas', *Philippine Daily Inquirer*, 28 August.

Caplan, P. (1987) *The Cultural Construction of Sexuality*, London: Tavistock Publications.

Cebu City Health Department (1992) 'Present situation', unpublished statistics, AIDS Prevention and Control Program/STD-AIDS Detection Center, Cebu City.

Chant, S. and McIliwaine, C. (1995) 'Gender and export manufacturing in the Philippines: continuity or change in female employment? The case of the Mactan Export Processing Zone', *Gender, Place and Culture* 2(2): 147–76.

Clifford, J. (1986) 'Introduction: partial truths', in J. Clifford and G. Marcus (eds) *Writing Culture: The Poetics and Politics of Ethnography*, Berkeley, CA: University of California Press, pp. 1–26.

—— (1988) *The Predicament of Culture: Twentieth Century Ethnography, Literature and Art*, Cambridge, MA: Harvard University Press.

Clifford, J. and Marcus, G. (eds) (1986) *Writing Culture: The Poetics and Politics of Ethnography*, Berkeley, CA: University of California Press.

Cohen, John M. and Uphoff, Norman T. (1980) 'Participation's place in rural development: seeking clarity through specificity', *World Development* 8: 213–35.

Conaco, C.G. (1992) 'Prostitutes and transmission of AIDS/HIV: implications on disease control management" *Philippine Journal of Public Administration* 36(3): 195–211.

Constable, N. (1997) *Maid to Order in Hong Kong: Stories of Filipina Workers*, Ithaca, NY and London: Cornell University Press.

Constantino-David, K. (1992) 'The Philippine experience in scaling-up', in M. Edwards and D. Hulme (eds) *Making a Difference: NGOs and Development in a Changing World*, London: Earthscan Publications, pp. 137–47.

Crimp, D. (ed.) (1989) *AIDS: Cultural Analysis, Cultural Activism*, Cambridge, MA: MIT Press.

—— (1992) 'Portraits of people with AIDS', in L. Grossberg, C. Nelson and P. Treichler (eds) *Cultural Studies*, London and New York: Routledge, pp. 117–33.

Crush, J. (1995) *Power of Development*, London and New York: Routledge.

Curtis, J. and Arreola, D. (1991) 'Zonas de tolerancia on the northern Mexican border', *The Geographical Review* 81(3): 333–46.

Dañguilan, M.J. (1993) *Making Choices in Good Faith: A Challenge to the Catholic Church's Teachings on Sexuality and Contraception*, Quezon City: WomanHealth.

David, R.D. (1991) 'Filipino workers in Japan: vulnerability and survival', *Kasarinlan* 6(3): 9–23.

de la Cerna, M. (1992) 'Women empowering women: the Cebu experience', *Review of Women's Studies* 3(1): 51–73.

Delacoste, F. and Alexander, P. (eds) (1987) *Sex Work: Writings by Women in the Sex Industry*, Pittsburgh: Cleis Press.

de Leon, A., Contor, E. and Abueva, A. (1991) *Tourism and Child Prostitution in the Philippines, Caught in Modern Slavery: Tourism and Child Prostitution in Asia*, Bangkok: The Ecumenical Coalition on Third World Tourism.

Desquitado, M. (1992) *Behind the Shadows: Toward a Better Understanding of Prostituted Women*, Davao: Talikala Inc.

De Stoop, C. (1994) *They are so Sweet, Sir: The Cruel World of Traffickers in Filipinas and Other Women*, Manila: Limitless Asia.

de Zalduondo, B.O. (1991) 'Prostitution viewed cross-culturally: toward recontextualising sex work in AIDS intervention research', *The Journal of Sex Research* 28(2): 223–48.

Dioneda, L. (1993) 'Women in the flesh industry', *Bakud* 3(2): 6–7.

Diprose, R. and Ferrell, R. (1991) *Cartographies: Poststructuralism and the Mapping of Bodies and Spaces*, Sydney: Allen and Unwin.

Doezema, J. (1998) 'Forced to choose: beyond the voluntary v. forced prostitution dichotomy', in K. Kempadoo and J. Doezema (eds) *Global Sex Workers: Rights, Resistance and Redefinition*, London and New York: Routledge, pp. 34–50.

DOH (Department of Health) AIDS Unit (1993) 'Report of the consultative meeting on injecting drug use and HIV in Cebu', unpublished report. Manila: National AIDS Prevention and Control Program.

Doyo, C.P. (1983) 'The prostitution problem must be viewed from a national, even global, perspective', *Philippine Panorama* 12: 36–40.

Duncan, J. and Ley, D. (eds) (1993) *Place/Culture/Representation*, New York and London: Routledge.

ECPAT (End Child Prostitution in Asian Tourism) (1994) *Tourism and Child Prostitution in Cebu*, Manila: ECPAT.

Enloe, C. (1983) *Does Khaki Become You? The Militarisation of Women's Lives*, London: Pluto Press.

—— (1989) *Bananas, Beaches and Bases: Making Feminist Sense of International Politics*, London: Pandora Press.

Enriquez, V. (1990) 'Indigenous personality theory', in V. Enriquez (ed.) *Indigenous Psychology*, Quezon City: Akademya ng Sikolohiyang Pilipino.

Escobar, A. (1995) *Encountering Development: The Making and Unmaking of the Third World*, Princeton, NJ: Princeton University Press.

Fee, E. and Fox, D.M. (eds) (1988) *AIDS: The Burdens of History*, Berkeley, CA: University of California Press.

Fernandez, L. (1993) 'Slaves', *Philippines Free Press*, 28 August, pp. 12, 49, 55, 78.

Foucault, M. (1977) *Discipline and Punish*, London: Penguin Books.

—— (1978) *The History of Sexuality: An Introduction*, Volume 1, London: Penguin Books.

—— (1980) 'Body/power', in Colin Gordon (ed.) *Power/knowledge: Selected interviews and other writings, 1972–1977*, New York: Pantheon Books, pp. 55–62.

Fuss, D. (1989) *Essentially Speaking: Feminism, Nature and Difference*, New York and London: Routledge.

Gallop, J. (1988) *Thinking through the Body*, New York: Columbia University Press.
Gatens, M. (1988) 'Towards a feminist philosophy of the body', in B. Caine, E. Grosz and M. de Lepervanche (eds) *Crossing Boundaries: Feminisms and the Critique of Knowledges*, Sydney: Allen and Unwin, pp. 59–70.
Gilman, S.L. (1988) *Disease and Representation: Images of Illness from Madness to AIDS*, Ithaca, NY: Cornell University Press.
Gorman, M. and Mallon, D. (1989) 'The role of a community-based health education program in the prevention of AIDS', *Medical Anthropology* 10: 159–66.
Gregory, D. (1994) *Geographical Imaginations*, Cambridge: Blackwell.
Grosz, E. (1989) *Sexual Subversions: Three French Feminists*, Sydney: Allen and Unwin.
—— (1993) 'Bodies and knowledges: feminism and the crisis of reason', in L. Alcoff and E. Potter (eds) *Feminist Epistemologies*, New York and London: Routledge, pp. 187–215.
—— (1994) *Volatile Bodies: Toward a Corporeal Feminism*, St Leonards, NSW: Allen and Unwin.
Grover, J.Z. (1992) 'AIDS, keywords and cultural work', in Grossberg, L. Nelson, C. and P. Treichler (eds) *Cultural Studies,* London and New York: Routledge, pp. 227–239.
Gwekoh, S.H. (1932) 'The city of evils,' *Progress*, 19 June.
Haraway, D. (1988) 'Situated knowledges: the science question in feminism and the privilege of partial perspective', *Feminist Studies* 14(3): 575–99.
—— (1990) 'A manifesto for cyborgs: science, technology, and socialist feminism in the 1980s', in L.J. Nicholson (ed.) *Feminism/Postmodernism*, New York and London: Routledge, pp. 190–233.
—— (1991) *Simians, Cyborgs and Women: The Reinvention of Nature*, London: Free Association Books.
Hartsock, N. (1990) 'Foucault on power: a theory for women?', in L.J. Nicholson (ed.), *Feminism/Postmodernism*, New York and London: Routledge, pp. 157–75.
Harvey, D. (1989) *The Condition of Postmodernity*, Cambridge: Blackwell.
Hochschild, A. R. (1983) *The Managed Heart: Commercialisation of Human Feeling*, Berkeley, CA: University of California Press.
hooks, b. (1991) *Yearning: Race, Gender, and Cultural Politics*, London: Turnaround.
Hubbard, P. (1997) 'Red-light districts and toleration zones: geographies of female street prostitution in England and Wales', *Area* 29(2): 129–40.
Irigaray, L. (1985) *This Sex Which is Not One*, transl. C. Porter with C. Burke, Ithaca, NY: Cornell University Press.
Jackson, P. (1991) 'The cultural politics of masculinity: towards a social geography', *Transactions of the Institute of British Geographers* 16: 199–213.
—— (1992) *Maps of Meaning*, New York and London: Routledge.
Jacobs, J. (1996) *Edge of Empire: Postcolonialism and the City*, London and New York: Routledge.
Jenness, V. (1993) *Making it Work: The Prostitutes' Rights Movement in Perspective*, New York: Aldine de Gruyter Inc.
Johnson, L. (1994) 'What future for feminist geography?', *Gender, Place and Culture* 1(1): 103–13.
Jolly, M. (1994) 'Introduction', in M. Jolly (ed.) 'Women's difference: Sexuality and maternity in colonial an postcolonial discourses', *The Australian Journal of Anthropology* 5(1 and 2): 1–10.
Jumilla, L.T. (1993) 'Brunei Beauties probed', *Philippine Daily Inquirer*, 5 August, pp. 1, 10.
Kane, S. and Mason, T. (1992) 'IV drug users and "sex partners": the limits of epidemiological categories and the ethnography of risk', in G. Herdt and

S. Lindenbaum (eds) *The Time of AIDS: Social Analysis, Theory and Method*, Newbury Park: Sage Publications.

Keith, M. (1992) 'Angry writing: (re)presenting the unethical world of the ethnographer', *Environment and Planning D: Society and Space* 10: 551–68.

Keith, M. and Pile, S. (eds) (1993) *Place and the Politics of Identity*, New York and London: Routledge.

Kempadoo, K. and Doezema, J. (eds) (1998) *Global Sex Workers: Rights, Resistance and Redefinition*, London and New York: Routledge.

Kondo, D. (1990) *Crafting Selves: Power, Gender and Discourses of Identity in a Japanese Workplace*, Chicago: University of Chicago Press.

Law, L. (1997) 'Cebu and *Ceboom*: the political place of globalisation in the Philippines', in P.J. Rimmer (ed.) *Pacific Rim Development: Integration and Globalisation in the Asia Pacific Economy*, Sydney: Allen and Unwin, pp. 240–66.

Laya Women's Collective (1993) 'Prostitution: old tale new face', *Laya Feminist Quarterly* 2(3): 1–56.

Lee, L. and WEDPRO (Women's Education, Development, Productivity and Research Organisation) (1992) *From Carriers to Communities: Alternative Employment, Economic Livelihood and Human Resource Development for Women in the Entertainment Industry*, Manila: WEDPRO.

Lewis, C. and Pile, S. (1996) 'Woman, body, space: Rio Carnival and the politics of performance', *Gender, Place and Culture* 3(1): 33–41.

Lewis, J. (1980) 'A boom Japan does not want', *Diliman Review* 28: 44–5.

Litong, G.T., Espartero, J., Pinchay, N., Goyena, M., Ursua, E. and S. Reyes (n.d.) 'Philippine laws and jurisprudence relevant to the AIDS/HIV problem', unpublished paper. Manila.

Lloyd, G. (1993) *The Man of Reason: 'Male' and 'Female' in Western philosophy*, London and New York: Routledge.

Longhurst, R. (1994) 'The geography closest in – the body . . . the politics of pregnability', *Australian Geographical Studies* 32(2): 214–23.

—— (1995) 'The body and geography', *Gender, Place and Culture* 2(1): 97–105.

Lyttleton, C. (1994) 'Knowledge and meaning: the AIDS education campaign in rural northeast Thailand', *Social Science and Medicine* 38(1): 135–46.

Mackie, V. (1988) 'Division of labour: multinational sex in Asia', in G. McCormack and Y. Sugimoto (eds) *The Japanese Trajectory: Modernisation and Beyond*, Cambridge: Cambridge University Press, pp.218–232.

McClintock, A. (1995) *Imperial Leather: Race, Gender and Sexuality in the Colonial Context*, London and New York: Routledge.

McKenzie, N.F. (1991) *The AIDS Reader: Social, Political and Ethical Issues*, New York: Penguin Group.

MacKinnon, C. (1987) *Feminism Unmodified: Discourses on Life and Law*, Cambridge, MA: Harvard University Press.

—— (1989) *Toward a Feminist Theory of the State*, Cambridge, MA: Harvard University Press.

Magno, A. R. (1993) 'A changed terrain for popular struggles', *Kasarinlan* 8(3): 6–21.

Malcomson, S. (1995) 'Disco dancing in Bulgaria', in M. Henderson (ed.) *Borders, Boundaries, and Frames: Essays in Cultural Criticism and Cultural Studies*, New York and London: Routledge.

Manderson, L. (1995) 'The pursuit of pleasure and the sale of sex', in P. Abramson (ed.) *Sexual Nature/Sexual Culture: Theorising Sexuality from the Perspective of Pleasure*, Chicago: University of Chicago Press.

Manderson, L. and Jolly, M. (eds) (1997) *Sites of Desire/Economies of Pleasure: Sexualities in Asia and the Pacific*, Chicago: University of Chicago Press.

Mangubat, E. (1993) 'Hooked on Nubain: Cebu drug addicts get high on injectable pain-killer', *Philippine Daily Inquirer*, 8 November.

Marchand, M.H. (1995) 'Latin American women speak on development: Are we listening yet?', in M.H. Marchand and J.L. Parpart (eds) *Feminism/Postmodernism/Development*, London and New York: Routledge, pp. 56–72.

Marchand, M.H. and Parpart, J.L. (1995) *Feminism/Postmodernism/Development*, London and New York: Routledge.

Marcus, G. and Fischer, M.M.J. (1986) *Anthropology as Cultural Critique: An Experimental Moment in the Human Sciences*, Chicago: University of Chicago Press.

Massey, D. (1994) *Space, Place and Gender*, Cambridge: Polity Press.

May, R.J. and Mathews, P. (1997) 'Forward', *Pilipinas* 29: iii–vi.

Miralao, V.A., Carlos, C.O. and Santos, A.F. (1990) *Women Entertainers in Angeles and Olongapo: A Survey Report*, Manila: Women's Education, Development, Productivity and Research Organization (WEDPRO) and Katipunan ng Kababaihan Para Sa Kalayaan (KALAYAAN).

Mohanty, C. (1988) 'Under western eyes: feminist scholarship and colonial discourses', *Feminist Review* 30: 61–88.

Mojares, R. (1993) 'Kamagayan: up from the swamps', *Sun Star Daily*, 12 May.

Moselina, L.M. (1979) 'Olongapo's rest and recreation industry: a sociological analysis of institutionalized prostitution with implications for a grassroots-oriented sociology', *Philippine Sociological Review* 27: 176–88.

Mortimer, L. (1994/95) 'Will the new woman keep some of the old organs?' *Arena Journal* New Series No. 4: 105–25.

Murray, A. (1995) 'Femme on the streets, butch in the sheets', in D. Bell and G. Valentine (eds) *Mapping Desire*, New York and London: Routledge.

—— (1996) 'Minding your peers and queers: female sex workers in the AIDS discourse in Australia and South-east Asia', *Gender, Place and Culture* 3(1): 43–59.

Nelson, R. (1985) 'Strategies for change, from the Australian Prostitutes Collective', in R. Perkins and G. Bennett (eds) *Being a Prostitute*, Sydney: Allen and Unwin, pp. 279–87.

Nguyet Erni, J. (1997) 'Of desire, the *farang*, and textual excursions: assembling "Asian AIDS"', *Cultural Studies* 11(1): 64–77.

Odzer, Cleo (1994) *Patpong Sisters: An American Woman's View of the Bangkok Sex World*, New York: Blue Moon Books.

Ong, A. (1987) *Spirits of Resistance and Capitalist Discipline: Factory Women in Malaysia*, Albany: State University of New York Press.

—— (1991) 'The gender and labour politics of postmodernity', *Annual Review of Anthropology* 20: 279–309.

Ortner, S. (1995) 'Resistance and the problem of ethnographic refusal', *Comparative Studies in Society and History* 37(1): 173–93.

Osteria, Trinidad S. and Okamura, Jonathan Y. (eds) (1986) *Participatory Approaches to Development: Experiences in the Philippines*, Manila: Diamond Jubilee Publication, De la Salle University.

Parpart, J.L. (1995) 'Deconstructing the development "expert": Gender, development and "vulnerable groups"', in M.H. Marchand and J.L. Parpart (eds) *Feminism/Postmodernism/Development*, London and New York: Routledge, pp. 221–43.

Pateman, C. (1988) *The Sexual Contract*, Cambridge: Polity Press.

Patton, C. (1990) *Inventing AIDS*, London and New York: Routledge.

Perkins, R. (1991) *Working Girls: Prostitutes, their Life and Social Control*, Canberra: Australian Institute of Criminology.

Pheterson, G. (ed.) (1989) *A Vindication of the Rights of Whores*, Seattle: Seal.

Pile, S. (1994) 'Masculinism, the use of dualistic epistemologies and third spaces', *Antipode* 26(3): 255–77.
—— (1997) 'Introduction', in S. Pile and M. Keith (eds) *Geographies of Resistance*, London and New York: Routledge, pp. 1–32.
Pile, S. and M. Keith (1997) *Geographies of Resistance*, London and New York: Routledge.
Pile, S. and N. Thrift (1995) *Mapping the Subject: Geographies of Cultural Transformation*, New York and London: Routledge.
Pratt, G. (1993) 'Reflections on postmodernism and feminist empirics, theory and practice', *Antipode*, 25(1): 51–63.
Probyn, E. (1991) 'This body which is not one: technologising an embodied self', *Hypatia* 6(3): 111–24.
Rafael, V. (1995) 'Introduction', in V. Rafael (ed.) *Discrepant Histories: Translocal Essays in Filipino Cultures*, Philadelphia: Temple University Press, pp. xiii–xxviii.
Richter, L.K. (1989) *The Politics of Tourism in Asia*, Honolulu: University of Hawaii Press.
Robinson, K. (1996) 'Of mail order brides and "boys' own" tales: representations of Asian-Australian marriages', *Feminist Review* 52: 53–68.
Rose, G. (1993a) 'Some notes towards thinking about the spaces of the future', in J. Bird, B. Curtis, T. Putnam, G. Robertson and L. Tickner (eds) *Mapping the Futures: Local Cultures, Global Changes*, New York and London: Routledge, pp. 70–83.
—— (1993b) *Feminism and Geography: The limits of Geographical Knowledge*, Minneapolis: University of Minnesota Press.
—— (1995) 'The interstitial perspective: a review essay on Homi Bhabha's *The location of culture*', *Environment and Planning D: Society and Space* 13: 365–73.
Rubin, G. (1984) 'Thinking sex: notes for a radical theory of the politics of sexuality', in C. Vance (ed.) *Pleasure and Danger: Exploring Female Sexuality*, London: Routledge and Kegan Paul, pp. 267–319.
Ryan, K. (1991) 'Sex, sin and work: changing perspectives on women, development and prostitution with a focus on Filipinas', MA thesis. The Flinders University of South Australia.
Said, E. (1991) *Orientalism*, London: Penguin.
Santos, A. (1992) 'Gathering the dust: the bases issue in the Philippines', in S. Pollock Sturdevant and B. Stoltzfus (eds) *Let the Good Times Roll: Prostitution and the US Military in Asia,* New York: The New Press, pp. 32–44.
Scott, J. (1985) *Weapons of the Weak: Everyday Forms of Peasant Resistance*, New Haven: Yale University Press.
Seeley, J., Kengeya-Kayondo, F. and Mulder, D. (1992) 'Community-based HIV/AIDS research – whither community participation? Unsolved problems in a research programmeme in rural Uganda', *Social Science and Medicine* 34(10): 1089–95.
Seidel, G. (1993) 'The competing discourses of HIV/AIDS in Sub-Saharan Africa: discourses of rights and control vs. discourses of control and exclusion', *Social Science and Medicine* 36(3): 175–94.
Slater, D. (1992) 'Theories of development and politics of the post-modern', *Development and Change* 23(3): 283–319.
Smith, A. and Kaminskas, G. (1992) 'Female Filipino migration to Australia: an overview', paper presented to the Fourth International Philippine Studies Conference, 1–3 July, Canberra.
Soja, E. (1989) *Postmodern Geographies: The Reassertion of Space in Critical Social Theory*, London: Verso.
Spivak, G.C. (1987) *In Other Worlds: Essays in Cultural Politics*, New York: Methuen.

—— (1988) 'Can the subaltern speak?', in G. Nelson and L. Grossberg (eds) *Marxism and the Interpretation of Culture*, Urbana: University of Illinois Press, pp. 271–313.

Stoler, A. (1995) *Race and the Education of Desire*, Durham, NC and London: Duke University Press.

Stoltzfus, B. and Sturdevant, S. (1990) 'From Samar to the Seventh Fleet: women's experiences of sexual labour in Olongapo', in D. Flamiano and D. Goertzen (eds) *Critical Decade: Prospects for Democracy in the Philippines in the 1990s*, Berkeley, CA: Philippine Resource Centre, pp. 137–42.

Sturdevant, S. Pollock and Stoltzfus, B. (1992) *Let the Good Times Roll:Prostitution and the US Military in Asia*, New York: The New Press.

Sullivan, B. (1994) Rethinking prostitution', paper presented to the Women's Studies Seminar Series, Australian National University, 20 May, Canberra.

Talikala (1993) 'A comprehensive survey on Davao City's entertainment industry and the hospitality women', unpublished report. Davao: Talikala Inc.

Tan, M., de Leon, A., Stoltzfus, B. and O'Donnell, C. (1989) 'AIDS as a political issue: working with the sexually prostituted in the Philippines', *Community Development Journal* 24(3): 186–93.

Tapales, P.D. (1988) 'The commercial chain of prostitution', *Philippine Currents* 3: 9–11.

—— (1990) 'Women, migration, and the mail-order bride phenomenon: focus on Australia', *Philippine Journal of Public Administration* 34(4): 311–22.

Treichler, P. (1989) 'Biomedical discourse: an epidemic of signification', in D. Crimp (ed.) *AIDS: Cultural Analysis/Cultural Activism*, Cambridge, MA: MIT Press, pp. 31–70.

Trumbach, R. (1991) 'Sex, gender, and sexual identity in modern culture: Male sodomy and female prostitution in enlightenment London', *Journal of the History of Sexuality* 2(2): 186–203.

Truong, T. (1990) *Sex, Money and Morality*, London: Zed Books Ltd.

Tyner, J.A. (1996) 'Constructions of Filipina migrant entertainers', *Gender, Place and Culture* 3(1): 77–93.

Valentine, G. (1989) 'The geography of women's fear', *Area* 21(4): 385–90.

—— (1993) '(Hetero)sexing space: lesbian perceptions and experiences of everyday spaces', *Environment and Planning D: Society and Space* 11: 395–413.

Vance, C.S. (1991) 'Anthropology rediscovers sexuality: A theoretical comment', *Social Science and Medicine* 33(8): 875–84.

VPHCS (Visayas Primary Health Care and Services) (n.d.) 'Survey reveals low knowledge on AIDS', *VPHCS Newsletter*, Cebu City.

Wall, D.R. (1983) 'Filipino brides: slaves or marriage partners? – A further comment', *Australian Journal of Social Issues* 18(1): 217–20.

Watney, S. (1989) 'The subject of AIDS', in P. Aggleton, G. Hart and P. Davies (eds) *AIDS: Social Representations, Social Practices*, Falmer Press, New York, pp. 64–73.

Watson, S. and Gibson, K. (eds) (1995) *Postmodern Cities and Spaces*, Oxford: Blackwell.

Weeks, J. (1985) *Sexuality and its Discontents: Meanings, Myths and Modern Sexualities*, London: Routledge and Kegan Paul.

—— (1987) 'Questions of identity', in P. Caplan (ed.) *The Cultural Construction of Sexuality*, London: Tavistock Publications, pp. 31–51.

White, L. (1990) *The Comforts of Home: Prostitution in Colonial Nairobi*, Chicago: University of Chicago Press.

WHO (World Health Organisation) (1989) *Collaboration with Non-governmental Organisations: The Development of Effective Cooperation at National, Regional, and International Levels*, Geneva: Global Program on AIDS.

Willis, P. (1977) *Learning to Labour: How Working Class Kids get Working Class Jobs*, Farnborough: Saxon House.

WRCC (Women's Resource Center of Cebu) (1993a) 'The dark side of Cebu's economic boom: women in the flesh industry', *Bakud* 3(2): 6–7.

—— (1993b) 'Prostitution', *Bakud* 3(2): 7, 14.

—— (1992) 'The women of Cebu: braving the odds', *Bakud*, 2(3): 5–11.

INDEX